Jimmy Alberto Aramendiz Fuentes

On the Influence of Piecewise Defined Contact Geometries on Friction Dampers

Karlsruher Institut für Technologie
Schriftenreihe des Instituts für Technische Mechanik

Band 40

Eine Übersicht aller bisher in dieser Schriftenreihe erschienenen Bände finden Sie am Ende des Buchs.

On the Influence of Piecewise Defined Contact Geometries on Friction Dampers

by
Jimmy Alberto Aramendiz Fuentes

KIT Scientific Publishing

Karlsruher Institut für Technologie
KIT-Fakultät für Maschinenbau

On the Influence of Piecewise Defined
Contact Geometries on Friction Dampers

Zur Erlangung des akademischen Grades eines Doktors der Ingenieurwissenschaften von der KIT-Fakultät für Maschinenbau des Karlsruher Instituts für Technologie (KIT) genehmigte Dissertation

von Jimmy Alberto Aramendiz Fuentes

Tag der mündlichen Prüfung: 15. September 2022
Hauptreferent: Prof. Dr.-Ing. habil. Alexander Fidlin
Korreferent: Prof. Dr.-Ing. Thomas Sattel

Impressum

KIT Scientific Publishing

Karlsruher Institut für Technologie (KIT)
KIT Scientific Publishing
Straße am Forum 2
D-76131 Karlsruhe

KIT Scientific Publishing is a registered trademark
of Karlsruhe Institute of Technology.
Reprint using the book cover is not allowed.

www.ksp.kit.edu

Print on Demand 2023 – Gedruckt auf FSC-zertifiziertem Papier

ISSN 1614-3914
ISBN 978-3-7315-1267-7
DOI 10.5445/KSP/1000153493

Kurzfassung

Die Reduktion von Schwingungen in Maschinen, Anlagen und Gebäuden ist eine wichtige Aufgabe im Ingenieurwesen. Vibrationen sind oft mit Geräuschen und Rauheit verbunden und sind im besten Fall nur für die Kundenzufriedenheit nachteilig. Im Worst Case führen sie zum Systemversagen. Im Gegensatz dazu führt Schwingungsminderung zu einer längeren Lebensdauer, sichereren Systemen und finanziellen Vorteilen durch weniger Ausfallzeiten. Daher ist es wichtig, effektive Dämpfer zu konstruieren.

Diese Arbeit betrachtet Dämpfer, die sich nicht auf eine Schwingungsreduktionsstrategie beschränken, sondern mehrere kombinieren, um optimale Ergebnisse zu erzielen. Die Möglichkeiten herkömmlicher Reibungsdämpfer werden durch stetige, stückweise definierten Kontaktgeometrien erweitert. Dies führt zu Reibungsdämpfern, die ihr Verhalten je nach Amplitude der Schwingungen ändern. Der passive, abgestimmte Keildämpfer wird entworfen und untersucht. Dieser Dämpfer bringt Dämpfung bei hohen Schwingungsamplituden in System ein und nutzt Tilgung bei niedrigen Schwingungsamplituden aus. Es werden numerische und analytische Untersuchungen durchgeführt. Um das qualitative Verhalten des Dämpfers zu validieren, wird ein Dämpferprototyp konstruiert und erprobt. Zudem wurde auch eine aktive Variante des abgestimmten Keildämpfers betrachtet. Es werden zwei Regelstrategien entworfen, die adaptive Mehrmodellregelung und die langsame, frequenzbasierte Regelung. Diese werden mit einer State-of-the-Art-Regelungsstrategie in transienten, quasistationären und Anwendungsszenarien verglichen.

Die Untersuchungen zum passiven, abgestimmten Keildämpfer zeigen, dass Dämpfung und Tilgung entkoppelt werden. Eine Optimierung der Dämpferparameter ergibt im Frequenzgang eine Reduktion der Maximalamplitude von 87.47 % unter Beibehaltung der Tilgung. Die Experimente validieren den Entkopplungseffekt sowie den qualitativen Einfluss der Parameter. Die aktiven Systeme erreichen mit Amplitudenabsenkungen von 91.11 % das beste Ergebnis.

Abstract

The reduction of vibrations in machines, plants, and buildings is a common task in engineering. Vibrations are often coupled with noise and harshness and are at least detrimental to customer satisfaction. In the worst case, they cause system failures. In contrast, the reduction of vibrations yields an increase in the life expectancy, safer systems, and financial gains due to less downtime. It is therefore essential to design devices that are capable of mitigating large vibrations.

This work considers dampers that do not solely focus on a single vibration reduction strategy but instead combine them to achieve optimal results. The dampers are based on dry friction dampers. However, the capabilities of conventional dry friction dampers are expanded by taking into account continuous piecewise defined contact geometries. This leads to friction dampers that change their behavior depending on the amplitude of the oscillations. The vibration damping device in this work, the tuned wedge damper, introduces damping at high oscillation amplitudes and takes advantage of absorption at low oscillation amplitudes. This passive system is investigated numerically and analytically. Additionally, a damper prototype is constructed and experiments are performed to validate the qualitative behavior of the damper. An active variant of the tuned wedge damper is also considered. Two novel control strategies are designed, namely the adaptive multiple model control and the slow frequency-based control. These are compared to a state-of-the-art control strategy for friction dampers in transient, quasistationary, and application scenarios.

The investigations into the passive design of the tuned wedge damper show that the damper decouples damping and absorption. An optimization of the damper parameters yields a maximum amplitude reduction in the system's frequency response function of 87.47% while maintaining an absorption frequency. The experiments validate the decoupling effect as well as the qualitative influence of parameter variations. Finally, the active systems achieve the best results with amplitude reductions of 91.11%.

Contents

Appendix

Vorwort

Die vorliegende Arbeit entstand während meiner Tätigkeit als wissenschaftlicher Mitarbeiter am Institut für Technische Mechanik, Bereich Dynamik/Mechatronik des Karlsruher Instituts für Technologie (KIT).

Ich möchte mich bei meinem Doktorvater Prof. Dr.-Ing. habil. Alexander Fidlin für seine wissenschaftliche Anleitung während meiner Promotionszeit. Ohne seine Ratschläge und Denkanstöße wäre meine Arbeit nicht in diese Form entstanden. Vor allem hat aber seine Betreuung mich zu einem besseren Ingenieur und einen Wissenschaftler mit hohem Standards gemacht. Auch die Freiheiten, die ich in meine Forschung hatte, haben mich Freude bereitet. Weiterhin möchte ich mich für alle Gespräche außerhalb des wissenschaftlichen Kontexts bedanken. Ich habe viel Geschichte und interessante Anmerkungen fürs Leben von Ihnen gelernt.

Für die Übernahme des Zweitgutachtens bedanke ich mich bei Prof. Dr.-Ing. Thomas Sattel vom Fachgebiet Mechatronik der Technischen Universität Ilmenau. Es freut mich, ein Korreferent zu haben, den ich schon länger kenne. Unsere Zusammenarbeit im Rahmen des DFG-SPP 1897 hat mich gefreut, und ich habe mich im Ilmenau sehr willkommen gefühlt. Für Ihre Anmerkungen im Rahmen des Promotionsverfahrens möchte ich mich auch bedanken, denn diese haben meine Arbeit bereichert.

Ich danke auch meine Kolleginnen und Kollegen, die mich am Institut für Technische Mechanik begleitet haben. Ich habe von Eurer Erfahrung und Denkanstöße profitieren können. Auch die nicht wissenschaftlichen Gespräche haben mich gefreut und entsprechend für Entspannung oder Ablenkung in stressigen Zeiten gesorgt. Besonders möchte ich Dr.-Ing. Jens Burgert hervorheben, mit dem ich mich ein Bürozimmer geteilt haben. Wir haben so viel Zeit miteinander verbracht und dafür bin ich dankbar. Es hat mich gefreut, dich morgens zu begrüßen und unsere gelegentlichen Überlegungen über Gott und die Welt. Ein besonderer Dank auch an Dr.-Ing. Ulrich Römer, der mich zuerst als Masterand den wissenschaftlichen Weg gezeigt hat. Als ich meine Zeit als wissenschaftlicher Mitarbeiter verbracht habe, hattest du oft gute Ratschläge zur Problemlösung, vielen Dank dafür! An die nicht namentlich genannten Kolleginnen und Kollegen möchte ausdrücklich noch mal mein Dank aussprechen. Ich habe mich wohlgefühlt und ich habe mich gefreut, mit Euch zusammenzuarbeiten. Auch wegen Euch kam ich gerne zur Arbeit.

Bei meinen Freunden und bei Familie Reumann möchte ich mich bedanken. Ihr habt mich außerhalb der wissenschaftlichen Arbeit unterstützt und für die notwendigen Ablenkungen gesorgt. Besonders möchte ich Lou-Ann Reumann hervorheben. Durch Ihre Unterstützung (und Geduld in schwierigen Zeiten) könnte ich diese Herausforderung meistern.

A mi familia en Colombia les doy las gracias por el sacrificio que han hecho por mí. A los Aramendiz y a los Fuentes les doy las gracias por acompañarme y recibirme tan bien en mis visitas anuales. A mi papá y mi mamá les agradezco la posibilidad de haber podido estudiar en Alemania. Entiendo que fue un sacrificio estar tan lejos, pero les agradezco esta oportunidad para salir adelante y desenvolver el potencial en mí. A mis hermanos les doy las gracias por acompañarme. El estar juntos, compartir y sus concejos me ha formado en parte como persona y por este aporte les doy las gracias.

Karlsruhe, den May 10, 2023
Jimmy Alberto Aramendiz Fuentes

1 Introduction

Even without our conscious perception, dry friction dampers are present in our everyday life and play a role in the technological advancement of mankind. They are found in the trains we use. They are essential in airplanes, specifically in turbines, used to cross thousands of kilometers in a matter of hours. They also help ensure the structural integrity of buildings. This work focuses on such dry friction dampers that help reduce vibrations in mechanical systems. In general, systems are not designed to withstand large vibration amplitudes. Prolonged exposure to large vibrations ultimately leads to failure and additional costs. Dry friction dampers use only the sliding contact between solid bodies to dissipate energy. The robustness and reliability of these dampers make them an appealing solution in various fields. Additionally, their natural ability to stick and slip offers advantages, which haven't been fully utilized. This work contributes to a better understanding of the behavior of such dampers. In this chapter, first, the motivation for the investigations is presented in section 1.1. Second, the state of research is detailed in section 2.1. Based on this literature overview the open research topics are identified and the specific purpose of this thesis is stated in section 1.3. The chapter concludes with the general thesis structure in section 1.4.

1.1 Motivation

Two main reasons drive the study of friction dampers and effective vibration reduction mechanisms in general: financial and environmental reasons. The implementation of effective dampers results in reduced vibration amplitudes and therefore lightweight and efficient machines. Such machines lead to better products and financial savings. Ultimately, they lead to a company's competitive edge. Additionally, vibration reduction also leads to a longer machine life and thus less unplanned downtime. In 2016, a study by the Wall Street Journal Custom Studios estimated the average cost of unplanned downtime for industrial manufacturers at $50 Billion per year with 42% due to equipment failure [181]. Environmental motives come from the energy efficiency challenges, which are required by politics. Therefore, there is a strong interest in the reduction of unnecessary energy costs in machines. The report of the International Energy Association "Net Zero by 2050" foresees a worldwide economic growth of 40% by 2030 with 7% less energy consumption than in 2020 [20]. Friction dampers must contribute toward the accomplishment of this goal. Specifically, they help to better understand friction's role in vibration reduction. In 1966 the costs due to the suboptimal utilization of friction were estimated at $200 billion [16]. The targeted study of friction dampers contributes to the efficient use of dry friction.

1.2 State of Research

This work investigates two types of friction dampers focusing on the geometrical design of the contact surfaces and their influence on vibration reduction. The specific investigation field of this work is a combination of two major fields: friction and vibration reduction. It is, therefore, only natural to first treat these two major fields separately and afterward consider their fusion, namely the field of friction dampers. This approach represents the structure of the state of research subchapters.

1.2.1 Literature Overview of Friction Research

A detailed review of the friction-induced vibration and friction modeling was made by Ibrahim [82, 83]. In the first part of the review, he detailed the different aspects and challenges that arise when modeling friction. In the second part, the review focused on the mechanisms that cause friction-induced vibrations and on examples in engineering applications. A detailed historical review was presented by Feeny et al. [48]. In this review, the contributions of friction towards technological advancements are made evident. Furthermore, an overview of the friction influence factors, fiction phenomena, and friction damping was given. Berger [14] gave a detailed report on the relationship between the system model and the friction model. He concluded that they cannot be chosen independently from one another and described the physical implications of the friction models. For the sake of brevity, only the most important aspects of these reviews are mentioned. Additionally, other sources found relevant by the author are included. In this chapter, first, the investigation into friction is introduced from a historical perspective. Second, the different factors that influence friction are presented. Third, the models that aim to describe such dependencies are detailed. Lastly, friction-induced phenomena are considered.

The first descriptions of friction in the context of mechanics are attributed to Leonardo Da Vinci. In his manuscripts, he noted the proportionality between the friction force and the normal load. These findings were lost because his work was not published [16]. However,

DA VINCI's results were independently rediscovered by GUILLAUME AMONTONS in 1699. He postulated two friction laws. The first one stated the proportional relationship between the friction force and the normal force. The second law postulated the independence of the friction force on the apparent contact area. The work of AMONTONS was verified and further expanded by CHARLES-AUGUSTIN COULOMB in 1785. He added the third law of friction, which states that the friction ratio is independent of the sliding velocity between the contact areas. Both AMONTONS and COULOMB were motivated by the better understanding and the reduction of friction in machines. As such, they experimented with wood, copper, iron, and steel material pairings in both dry and smeared configurations [35]. These three statements make up the AMONTONS-COULOMB friction laws and built the basis of macrotribology. LEONHARD EULER also made a substantial contribution in 1750, since he was the first to introduce a friction coefficient and a difference between static and dynamic friction. His work was theoretical in nature and it stated both the letter μ as the friction coefficient and the fundamental geometrical interpretation of the friction cone. Both these concepts are still used today. Although not explicitly mentioned in the friction laws, COULOMB and EULER had an intuitive understanding that the ratio between the normal and friction force was dependent on the material pairing and lubrication. A modern calculation of these ratios expressed as the friction coefficient was presented in the work of MAISSEN [115].

The aforementioned concepts laid a foundation for the study of friction. However, the state of research shows that friction characterization is not that simple. Since the AMONTONS-COULOMB friction laws, additional influence factors of friction have been identified. This is mainly due to new technological applications and advances. For example, in contrast to AMONTON's friction laws BODEN and TABOR [23] postulated that friction is dependent on the true contact area. Different from the apparent contact area, the true contact area is given by the sum of the junctions between the asperities that are actually in contact with each other. Furthermore, the true contact area is itself dependent on the surface roughness and the normal load. Both these factors influence the junctions between the asperities in the contact area. Depending on the normal load the asperities will deform either elastically, plastically, or even break. The sum of these behaviors determines the true contact area. Research into this relationship was further expanded by ANDREW et al. [4], TABOR [167], and SAKAMOTO [148]. Models, that approximate the contact area, were proposed by GREENWOOD and WILLIAMSON [68] and WHITEHOUSE and ARCHARD [179]. These works assume a Gaussian distribution of asperity and demonstrated that the true contact area depends on the normal load and helped to describe the underlying contact mechanics.

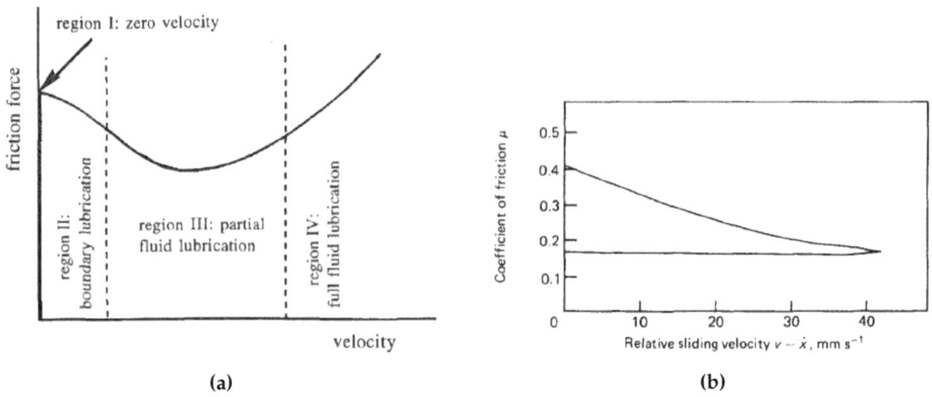

Figure 1.1: (a) Exemplary Stribeck curve, source: [82]. (b) Acceleration influence on the friction coefficient, source: [148].

Modern research has also corrected Coulomb's friction law. One significant work was derived by Stribeck [163]. His work focused on lubricated bearings and included four different frictions regimes depending on the sliding velocity: static friction, boundary lubrication, partial fluid lubrication, and full fluid lubrication, see Fig. 1.1a. However, Stribeck's studies do not imply a corresponding relation between the velocity and the friction coefficient. Sampson et al. [149] measured the effect of acceleration and deceleration on the friction force and noticed different values. Their measurements revealed a declining friction coefficient for increasing velocities and a constant friction value for declining velocities. These results were verified by Ko and Brockley [97] and Sakamoto [148], see Fig. 1.1b. Ko and Brockley [97] also noticed a humped friction force curve, which showed a local maximum and afterward a minimum for progressing velocities. Gao and Kuhlmann-Wilsdorf [63] found similar results for dry friction contacts in vacuum. However, they measured first a local minimum and afterward a local maximum for low speeds before the friction coefficient became independent of the velocity. Additional velocity curves for the friction coefficients were measured by Vinogradov [174], Grosch [70], Krauter [101], Martins [119], and Kapelke [89].

The static friction force is also dependent on the rate of compression and the time of stationary contact as suggested by Rabinowicz [142], Brockley [24], and Martins [119]. Both of these influence factors ultimately represent a dependency of friction on time. The two factors contribute to the shear strength of the asperities in the contact area. The normal load creates new junctions between the asperities, whereas the stationary time strengthens the bonds between asperities, i.e. the cold-welding effect. Plint and Plint also noticed this behavior in their experiments with smeared contacts [138]. However, they attributed this to the squeeze-film effect. Although the effect was different, the underlying result was the same, namely strengthening of the bonds between asperities due to the passage of time. Aronov et al. [7] and Dweib and D'Souza [42] also carried out experiments that determined that the friction force depends on the normal load for a constant sliding velocity, see Fig. 1.2a. The results of this research determined four regimes for rising normal loads. The first is a steady-state region where the friction is proportional to the load. Second, comes a nonlinear regime where the friction coefficient is not constant. Third a transient regime in which the friction fluctuates, and lastly, a self-excited regime where instabilities are observed.

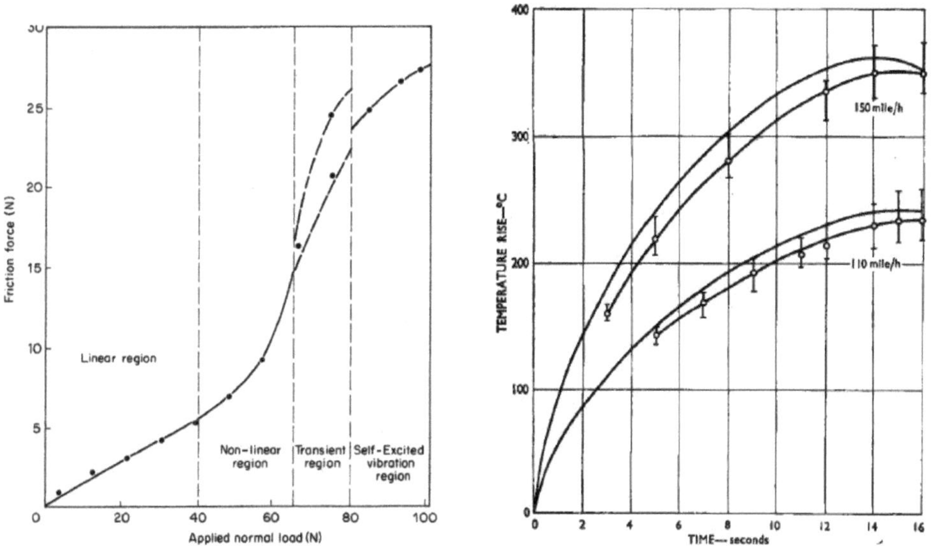

(a) (b)

Figure 1.2: (a) Normal load influence on friction force, source: [42]. (b) Temperature rise in friction for different sliding velocities, source: [130].

An additional influence factor in the description of friction is the temperature of the contact surface. It results from the heat generated due to friction. Bowden and Tabor [21] and Sarkar [151] published in their books theoretical and experimental results regarding effects of heat transfer on friction. They measured the well-observed phenomenon of rising contact temperature with rising contact velocities. Additionally, they showed that some material pairings show a saturation of the contact temperature. Similar observations were made by Newcomb [130], see Fig. 1.2b. The heat influence on friction was considered locally by Bhusan [15] and Kuhlmann-Wilsdorf [103], whereas Maksimov [116] assumed a uniformly distributed temperature of the contact surfaces. Their research offered insights into the instability mechanism that arises from the energy exchange between mechanical and thermal modes.

The overview above gives an insight into the most influential factors on friction. Nevertheless, additional influence factors exist such as the local wear of the contact surfaces studied by Bowden and Leben [22], the contact stiffness investigated by Andrew et al. [4], and the effects of dynamic normal or tangential forces acting on the contact surface as indicated by Godfrey [67] and Keer et al. [92]. Although friction has a variety of influence factors, this has not kept researchers from proposing models to describe certain aspects of these dependencies.

Some models focus on the microscopical contact mechanics, which occur between asperities in the contact surface as in the works of Greenwood and Williamson [68], Hisikado [79], and Tabor [167]. Partly, these works stated that the friction force is composed of two portions: one which is attributed to shearing and the other to plowing. The shearing portion of the friction force is attributed to the van der Waals forces between the atoms at the asperity junctions. The plowing part of the friction force represents the force necessary to move aside the material in the path of the penetrating asperities. This second friction mechanism becomes more relevant when one material is significantly harder than the other one. If this is not the case, the shear portion is dominant.

A large number of friction models are phenomenological and focus on the macroscopic level. The most common model is the Coulomb dry friction. While sticking the relative velocity is zero and friction force takes on the necessary value to uphold the corresponding sum of forces. This case holds as long as the friction force does not exceed the maximum value given by the stiction force $H = \mu N$. In the sliding case, the Amontons-Coulomb friction laws are expressed in the equation

$$R = \mu N \operatorname{sgn}(v_{\text{rel}})\,.$$

Where R is the friction force, μ is the friction coefficient, N is the normal load, v_{rel} is the relative velocity of the contact surfaces, and $\operatorname{sgn}(\cdot)$ represents the sign function

$$\operatorname{sgn}(x) := \begin{cases} -1, & x < 0 \\ 0, & x = 0 \\ 1, & x > 0 \end{cases}\,.$$

Although it is known that this is not the most accurate description of friction, this model has proven to yield acceptable results in the description of friction-induced vibrations. Additionally, for a plastic material behavior, Bowden and Tabor motivated the proportionality between friction and the normal load with microscopical contact mechanics. Assuming the dry contact of two rough surfaces with similar hardness, the friction coefficient is interpreted as the ratio between the shear strength to the yield stress [167]. With A_0 as the true contact area, τ_0 as the shear strength, and σ_0 as the yield stress this interpretation is summarized the equation

$$R = A_0\tau_0 = \frac{N}{\sigma_0}\tau_0 \overset{!}{=} \mu N \quad \rightarrow \quad \mu = \frac{\tau_0}{\sigma_0} \, .$$

A similar results for an elastic material behavior was provided by GREENWOOD and WILLIAMSON [68]. Therefore it is noted that under the assumed conditions the COULOMB friction model offers an acceptable approximation provided the friction coefficient μ is chosen accordingly. Variations and extensions of this model have been proposed to take into account additional factors. One extension is the consideration of different static and dynamic friction coefficients. For example, a velocity-dependent friction coefficient $\mu(v_{\text{rel}})$ was used by THOMSEN and FIDLIN [168]. This allows the consideration of the STRIBECK effect and is found in different variations as summarized by ARMSTRONG et al. [6].

In an effort to reduce computational effort, a common modification of the COULOMB model is the regularization of the sign function [132]. The two most common alternatives are

$$\operatorname{sgn}(x) \approx \tanh\left(\frac{x}{\nu}\right) \quad \text{and} \quad \operatorname{sgn}(x) \approx \frac{2}{\pi}\arctan\left(\frac{x}{\nu}\right) ,$$

where ν represents the regularization factor. As $\nu \rightarrow 0$ the curves tend to the sign function, however, stiction is impossible in such models. An alternative approach to reduce the computational effort was presented by KARNOPP [90]. He defined a small velocity region in the vicinity of $v_{\text{rel}} = 0$ where sticking is allowed even though strictly $v_{\text{rel}} \neq 0$. This avoids the computational effort associated with event detection.

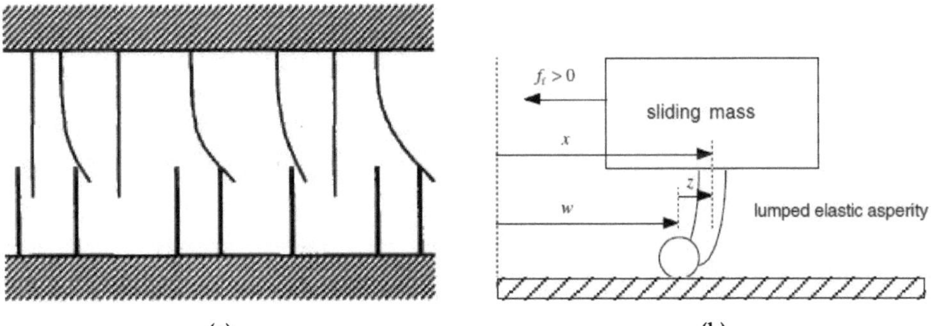

(a) (b)

Figure 1.3: (a) Multiple bristle model visualization, source: [26]. (b) Lumped bristle model visualization, source: [39].

The models mentioned above are algebraic models of friction. Therefore, they are not able to take into account the history of the contact and depend exclusively on the current state of the system. DAHL [31] formulated a friction model in which the friction force was described with a differential equation. This allowed the model to take into account the history of the contact and implement a model with memory. DAHL's model additionally introduced contact compliance which occurs when the stiction force is not exceeded. Similar friction models based on a bristle interpretation of friction were formulated by HAESSIG and FRIEDLAND [73], see Fig. 1.3a. This approach models the asperities deformation in the contact region as bristle interactions. To this end, additional variables and differential equations are introduced. Although not originally formulated as a bristle model the DAHL model is often interpreted as such. An expansion of these ideas was formulated by CANUDAS DE WIT et al. in the LuGre model [26], which is named after its origin in the cities of Lund and Grenoble. The LuGre model takes into account velocity-dependent friction coefficients, which the DAHL model omitted. Further advancement of bristle models is represented by the elastoplastic friction model proposed by DUPONT et al. [39, 40], see Fig. 1.3b. This model accurately simulates sticking in comparison to the DAHL and LuGre models, which show a drift when the forces acting on the friction are lower than the stiction force. For the stationary case, the DAHL, LuGre, and elastoplastic models all converge to the COULOMB friction model [89].

All the models mentioned above take their chosen representation of friction and solve the corresponding differential equations. Taking into account that while sticking the friction force takes on multiple values depending on the acting forces, a set-valued interpretation of the friction force is possible. This interpretation leads to differential inclusions and adds a degree of difficulty to the system solution. For example, the solution of such systems is no longer guaranteed. Advances into the mathematical solutions of such systems were contributed by FILIPPOV [57]. An analysis of differential inclusions with dry friction and possible bifurcations was presented by LEINE [107].

A variety of models for the simulation of friction exist. For example, the discrete asperities model proposed by JENKIN and EWIG modeled asperities with the JENKIN element, a serial combination of a spring and a dry friction element [38]. Another example is MINDLIN's work. He took the HERTZ contact mechanics expanded this work by taking into account a tangential load [123]. There are more friction models. However, for the sake of brevity, the author limits himself to friction models described above.

Taking into account the described friction influence factors and using appropriate models researchers take on the task of describing friction-induced vibrations. If left unchecked these oscillations cause discomfort, wear, and in worst-case system failure.

Den Hartog analyzed a case in which such vibrations caused the system failure of a drawbridge [33]. Two major mechanisms are attributed to friction-induced vibrations: the negative friction gradient and the modal coupling with nonconservative forces. Such oscillations may lead to quasi-harmonic oscillations as shown by Brockley and Ko [25] but may also lead to chaos as studied by Popp and Stelter [139, 141].

An explanation of the negative friction gradient mechanism was proposed by Blok [18]. He linearized the friction velocity curve and described the self-excitation linked to the negative gradient in the friction velocity curve. At low velocities, the negative slope causes friction to acts as an energy input which ultimately leads to instability. He also found that increased linear damping quenches these oscillations. Depending on the velocity of the oscillations, this self-excitation mechanism leads either to a pure-slip behavior or to stick-slip oscillations. Low oscillation velocities yield pure-slip oscillations, whereas oscillations with large enough velocities fulfill the stiction conditions and result in stick-slip oscillations. Additional works concerning self-excited oscillations due to a negative friction gradient were made by Hetzler [77, 78] and Thomsen and Fidlin [168].

The second mechanism is commonly referred to as flutter and is caused by modal coupling due to nonconservative friction forces. A review of the work into the instability mechanism was presented by Kinkaid et al. [96]. A minimal model that explained the instability sources of flutter was given by Hoffmann et al. [80], see Fig. 1.4a. Minimal models in the context of the automotive branch were investigated by von Wagner [175]. Hervé et al. [76], and Fidlin et al. [52], see Fig. 1.4b. The description of this mechanism leads to the explanation of friction-induced vibrations at high frequencies [77].

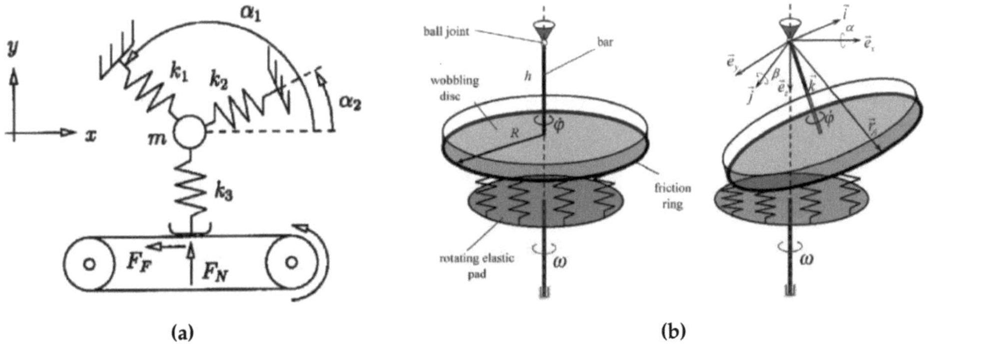

Figure 1.4: (a) A minimal model for flutter instability, source: [80]. (b) A minimal model for flutter instability in disks, source: [52].

Friction-induced oscillations are found in various engineering applications such as lubricated bearings, wheel/rail systems, brakes, and machining. The vibrations are often audible and in some cases even desired, e.g. string instruments [16, 48]. However, for the most part, these audible vibrations lead to discomfort. Ibrahim grouped these phenomena into two groups: chatter and squeal. Chatter occurs at low frequencies, whereas squeal occurs at higher frequencies. In the rotor dynamics field, friction-induced oscillations occur at low rotor velocities, when the rotation speed is not able to separate the rotor from the bearing. In such cases, dry friction contact dominates the contact dynamics between rotor and bearing. Smith and Pam experimented with these systems and concluded that such oscillations are caused either by increasing the load at a constant speed or by reducing the speed at a constant load [157]. Childs also studied such interactions and found that they lead to a parametric excitation in rotors [28]. Squeal in connection to dry friction and wheel-rail interactions occurs when trains take curves with a short radius. Bender and Remmington investigated such friction-induced vibrations [13] and also Scheider et al. [153]. The work of Remmington [143] showed that such oscillations are caused by a lateral creep of the wheels on the rails. In the context of disc brakes, the mechanisms that lead to brake noise are the negative friction gradient, sprag-slip, and flutter. The effect of a negative velocity gradient in the context of brakes was described by Fosberry and Holubecki [58]. The sprag-slip mechanism was proposed by Spurr [160] and is caused by a variation of the normal load, which leads to intermittent oscillations with slip and stick regimes. This mechanism differs substantially from the typical stick-slip oscillations since the latter occurs for constant normal loads. Investigations into flutter and disk brakes were made by Kinkaid et al. [96] and Fidlin and Stamm [55]. In machining, friction-induced vibrations lead to poor processing quality and are caused mainly by a negative friction gradient. Investigations and minimal models concerning friction in machining are presented by Tobias [170], Tlusty [169], and Moon [126].

As summarized above, friction is an intricate field with various influence factors. Different models are proposed to explain some aspects of these. However, there exists no valid and general model to take into account all of its influence factors. Friction serves to dissipate energy, however, under some circumstances it also induces vibrations. These vibrations are either desired as in the case of string instruments or undesired as in the case of brake squeal. The quenching of unwanted friction-induced vibrations and vibrations, in general, is another relevant field for engineers.

1.2.2 Literature Overview of Vibration Reduction Research

Investigations into vibration reduction began more than 100 years ago before Frahm filed a patent for the first vibration reduction device in October 1909 [59]. Valuable mechanical insights into the field of vibration reduction were made by Den Hartog in the 1930s [33]. He studied vibration reduction mechanisms in linear configurations with varying degrees of freedom. Such mechanisms rely mainly on either: vibration absorption, vibration damping, or vibration isolation. Absorbers use a auxiliary mass that is connected via an auxiliary spring to the main system. The parameters of the absorber are chosen so that its resulting movement counteracts the forces acting on the main system. Dampers are vibration reduction devices that dissipate mechanical energy from the system by converting it into another type, e.g. heat. Isolators are devices that are placed between the main system and the vibration source. Their task is to transmit as few vibrations as possible to the main system. Tuned mass dampers combine a vibration absorber with a dissipative element that is placed in parallel to the auxiliary spring. Depending on how the parameters are selected, it is absorption focused or damping focused. Since for all practical purposes material damping is present in every spring, the term vibration absorber is often used in the literature to indicate a lightly damped tuned mass damper. There is certainly no lack of creativity when designing tuned mass dampers, as shown in a survey by Sun et al. [164]. In this work passive, adaptive, and active variations of the linear tuned mass damper were considered. A review of tuned mass dampers in the context of structural engineering was presented by Gutierrez Soto and Adeli [71] and by Elias and Matsagar [45]. A review of nonlinear vibration dampers is given by Lu et al. [113]. A general way of classifying vibration reduction mechanisms is either by their linearity (linear or nonlinear systems) or by their energy source (passive or active systems). The combination of these two classifying qualities yields four categories: passive linear mechanisms, active linear mechanisms, passive nonlinear mechanisms, and active nonlinear mechanisms. This state of research subchapter is presented according to these categories.

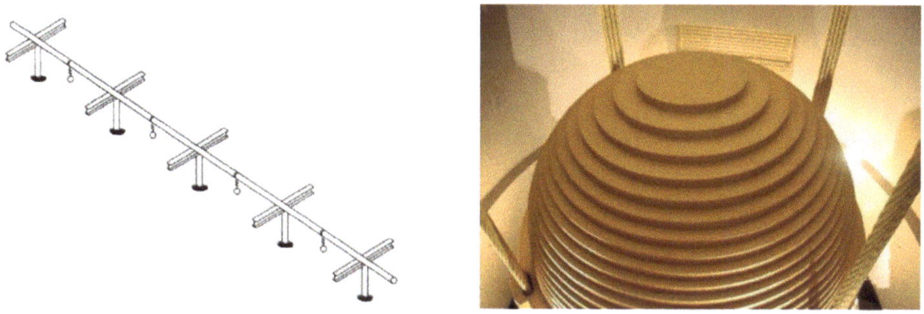

(a) (b)

Figure 1.5: (a) A mass-spring tuned mass damper for arctic pipelines, source: [74]. (b) Vibration absorber in the Taipei 101, source: [176].

The simplest implementation of a tuned mass damper is the conventional spring-mass system, as shown in Frahm's original patent [59]. However, this variant still proves effective in modern applications. HART et al. used this simple variant to reduce vibrations in arctic pipelines [74], see Fig. 1.5a. More involved implementations have been realized, such as the beam-type tuned mass damper proposed by AIDA et al. [1] or the tuned mass damper with cables found in the Taipei 101 [176], see Fig. 1.5b. Described by linear differential equations, the solution, handling, and design of these devices were thoroughly investigated in the first half of the 20th century. Such systems are designed to optimize either the vibration reduction at a single oscillation frequency or a desired frequency range. To this end, either the whole system is considered [33] or only the driving point, where the device is attached [46]. The latter constitutes the impedance coupling method proposed by EWINS. The single frequency optimization yields high peaks at the structural resonance frequencies of the system. If a broadband approach is required the equal peak method offers the best possible results. In the literature mainly two variations of this method are considered and were proposed by DEN HARTOG [33] and SNOWDON [158]. Both these works offer approximations, to achieve equal peaks at the system's structural resonance regimes. A closed-form solution to this problem was presented recently by ASAMI and NISHIHARA [8]. The different goals of these design strategies represent the limitation of the classical tuned mass damper: an optimization of both the structural resonances and the tuned frequency is not possible [33]. To further improve the results of FRAHM's original proposal researchers consider active and nonlinear variations of the tuned mass damper.

The next natural step in the improvement of such systems is the study of active vibration reduction mechanisms. Advances in this field started in the 1950s [91], and practical applications were realized in the late 1960s in the aerospace field by SMITH and LUM [156] and SCHUBERT and RUZICKA [154]. Since then, these concepts have only been developed further. Depending on their design these systems are classified into fully active, hybrid, or semi-active [164]. Fully active systems generate a force that acts on the oscillating structure and have an input-affine structure. Thus, their solution is calculated with manageable effort. However, this is normally coupled with high energy costs. An active implementation of a vibration damper was presented by BAILEY and HUBBARD [11]. In their work, they used a distributed piezoelectric actuator and a Ljapunov controller to reduce vibration oscillations in a Cantilever Beam. KIM et al. made similar investigations in the damping of beam vibrations [95]. However, they used a local piezoelectric actuator and an active dynamic vibration absorber controller to suppress structural resonances, see Fig. 1.6a. The hybrid vibration absorbers work in a partially active and a partially passive way depending on the external conditions and are often found in seismic structures. FUJITA used the active modes of the hybrid damper to reduce vibrations due to winds and weak earthquakes, whereas the passive mode was used to control vibrations caused by strong earthquakes. The proposed system was implemented in the

Long Term Credit Bank of Japan [61]. Additionally, LEE-GLAUSER et al. found that hybrid dampers were effective in the vibration reduction of broadband earthquakes [106]. On the other hand, semi-active mechanisms change a parameter, e.g. a damping or stiffness coefficient, to change the force acting on the oscillating structure. However, this indirect change of the force has its disadvantages. Since the controlled parameter often multiplies a state variable of the system, the control design is nonlinear. This added design effort is one of its main disadvantages. Furthermore, not every desired force can be generated, due to the parameter range limitation, e.g. positive parameters. Nevertheless, a semi-active implementation also leads to lower energy costs, making this an appealing solution. Such vibration reduction mechanisms are found in the work of KARNOPP, where the damping coefficient of an isolator was switched on and off dependent on the velocities of the system's structure [91]. DAVIS and LESIEUTRE proposed a semi-active vibration absorber that changed the stiffness of a piezoceramic element electrically [32]. The stiffness change allowed the vibration absorber to adapt its absorption frequency to the system's oscillations and effectively make use of this vibration reduction mechanism.

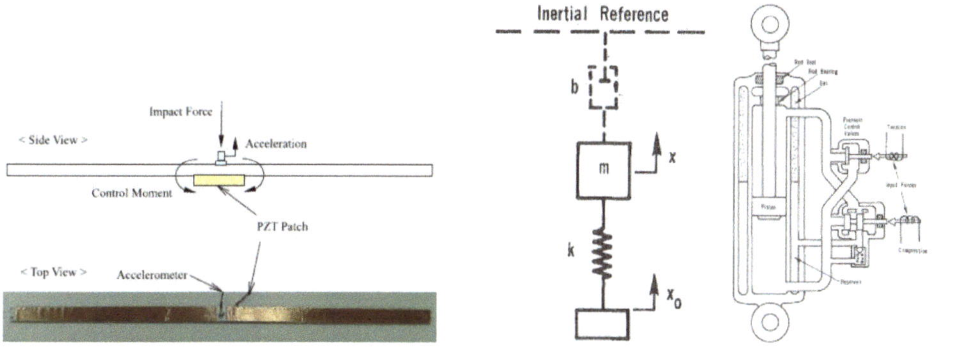

(a) **(b)**

Figure 1.6: (a) An active vibration absorber for beam vibration reduction, source: [95]. (b) An ideal and schematic depiction of the Skyhook damper, source: [91].

Additional design effort is required by active vibration dampers due to the design of the control strategy. An essential contribution was made by KARNOPP with the development of the Skyhook Control Strategy. This strategy was used for vibration isolation and was often implemented in the automotive branch, to isolate automobiles from ground vibrations. The main idea behind this approach is the semi-active control of a damping element that tries to emulate a damper connected to an inertial reference frame, cf. Fig. 1.6b. To this end, the damper dissipates energy whenever possible, and when this is not the case it does not exacerbate the vibrations of the considered system. Variations of this control strategy were investigated by LIU et al. [109]. An overview of additional control strategies was made by Alkhatib and Golnaraghi [2]. In their review, they considered feedforward, adaptive, coordinate coupling, robust, and optimal control. A review of vibration control strategies in the context of offshore structures was presented by Kandasamy [88].

An alternative way to improve the performance of passive linear vibration reduction mechanisms is the consideration of nonlinear effects. Although active linear dampers offer an improvement, they are bound by the limitations of linear systems. Passive nonlinear devices offer improvements over linear systems albeit they do require a higher design effort. A review of nonlinear passive elements was presented by Lu et al. [113]. Nonlinear vibration reduction devices were categorized depending on the placement of their nonlinearity. The nonlinearities appears on the stiffness, the damping, or in both stiffness and damping terms. This classification is adopted in this work. The categories are mainly represented by three devices: nonlinear energy sinks, nonlinear viscous dampers, and vibro-impact dampers.

Nonlinear energy sinks are a realization of nonlinear dissipative devices with nonlinear stiffness and are composed of three components: an auxiliary mass, a strong nonlinear stiffness, and a linear damping element see Fig. 1.7a. The nonlinear restoring force transfers vibrations from the main structure to the nonlinear energy sink. Afterward, the energy within the nonlinear energy sink is dissipated via the damping element. A review focused on nonlinear energy sinks and targeted energy transfer was given by Lee et al. [106]. Noticeable contributions in this field were made by Vakakis mainly in combination with cubic spring as a nonlinear stiffness [173]. Habib et al. and Detroux et al. noted a generalization of Den Hartog's equal peak method for nonlinear systems [72] and contributed to the design and analysis of such devices [36]. Although not explicitly

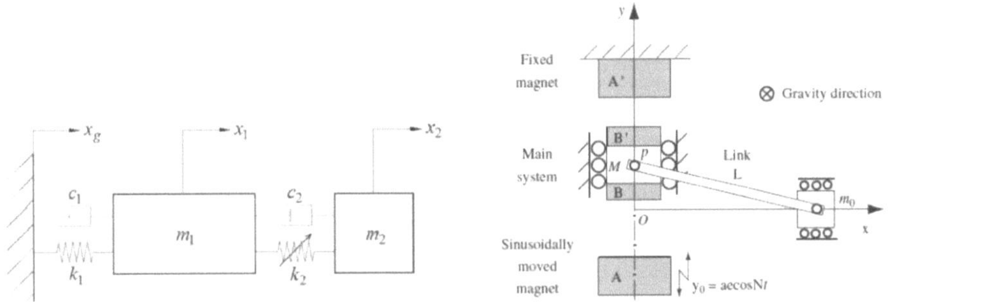

Figure 1.7: (a) An schematic depiction of a nonlinear energy sink, source: [113]. (b) System considered by Jo and Yabuno source: [86].

named a nonlinear energy sink, Jo and Yabuno considered the similar effect in a system with linear damping and quadratic and cubic stiffnesses generated by geometric configuration with levers [86], see Fig. 1.7b. This configuration reduced parametric excitations in the considered system.

A detailed work of nonlinear viscous dampers was made by Symans and Constantinou [166] in the context of structural engineering. These nonlinear devices use a viscous fluid, which is forced to flow through orifices or chambers, and have a damping force that depends only on the velocity. The resulting force is in general expressed as

$$F_d = d\text{sgn}(v)|v|^{\alpha}\ .$$

Where d represents the damping coefficient, v the velocity, and α the velocity exponent that defines the nonlinear characteristic. Rüdinger found advantages of such dampers in systems excited by random white noise excitation [146]. Martinez-Rodrigo and Romero focused on optimal strategies in the post-construction fitting of linear and nonlinear viscous dampers in buildings. Nonlinear devices reduced vibrations 35% more than their linear counterparts [118]. Advantages concerning the risk assessment and vibration reduction in seismic structures were presented by Tubaldi et al. [172].

Devices with nonlinear stiffness and nonlinear damping are constructed in part by the combination of the aforementioned mechanisms. Starosvetsky and Gendelman considered a cubic spring, however, in combination with quadratic damping [162]. The work focused on the targeted energy transfers mechanism and combined it with the advantages of nonlinear damping. On the other hand, some systems have the combination of nonlinear stiffness and damping intertwined in the system's design. This is the case for vibro-impact dampers and tuned liquid dampers. Vibro-impact dampers, also commonly known as particle impact dampers, are composed of small particles inside one or more collision chambers. The collision chamber is attached to the main structure, and the energy of the main structure's vibrations is dissipated into heat due to friction and partially inelastic collisions. Valuable contributions in this field date back to the 1960s by Masri. His work studied a single impact mass and a single impact chamber [121]. Studies into single impact masses in multiple impact chambers were conducted by Bapat and Sankar [12]. A particle impact damper with multiple small particles in a single collision chamber was investigated by Marhadi and Kinra [117], see Fig. 1.8a. Panossian highlighted that this damping mechanism was implemented in a non-obtrusive way into a space shuttle main engine liquid oxygen inlet tee [135], see Fig. 1.8b. Additional overviews of vibro-impact systems are presented by Babitsky [10] and Lu [112]. Liquid tuned dampers function similarly to vibration impact dampers,

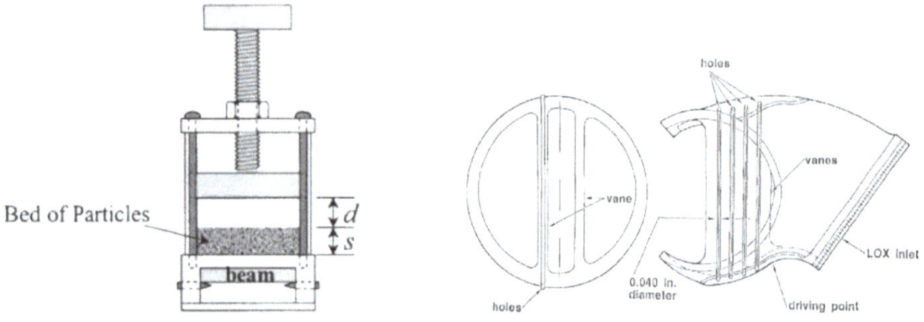

(a) (b)

Figure 1.8: (a) Frontal view of a particle impact damper on a cantilever beam, source: [117]. (b) A space shuttle main engine liquid oxygen inlet tee with holes for particles, source: [135].

however, the chamber is filled with a liquid instead of particles. Although there were other vibration reduction devices using liquids before, the term tuned liquid damper was introduced by Fujino et al. [60]. In contrast to earlier publications, their work proposed a shallow filled container which allowed higher damping. Modi and Munshi considered a tuned liquid damper with obstacles in the tank that lead to increased energy dissipation [125], see Fig. 1.9a. Multiple liquid tuned dampers attached to the same structure were investigated by Love and Tait [110]. An equivalent model for such dampers composed of a vibro-impact damper with a single particle attached via springs to its impact chamber, was developed by Farid and Gendelman [47], see Fig. 1.9b.

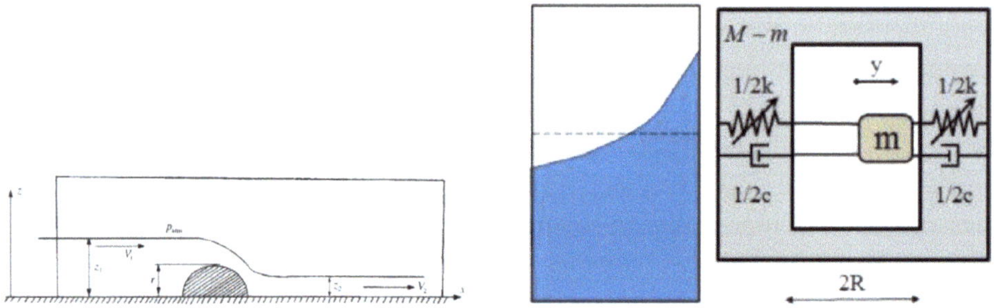

(a) (b)

Figure 1.9: (a) A tuned liquid damper with obstacles, source: [125]. (b) An equivalent model for the tuned liquid damper, source: [47].

Active nonlinear dampers combine the advantages of active control and nonlinear dampers. However, they have the highest design effort. Since the whole system is nonlinear, advanced control strategies are required. Intensely investigated nonlinear semi-active dampers are magneto- or electro-rheological dampers. These devices apply a magnetic/electric field to change the properties of a non-Newtonian fluid. Such dampers are represented with a BINGHAM model which is a parallel configuration of a linear spring, a viscous damper, and a dry-friction element [165]. The spring and damper model the fluid's elasticity and viscosity, whereas the dry-friction element models the building of chains and locking of active particles in the fluid. Depending on the magnitude of the applied field the break-away force of the friction element is varied. DYKE et al. considered a magneto-rheological damper combined with a clipped optimal control algorithm in building structures [43]. SUN and THOMAS studied an electro-rheological damper with an on-off control strategy, to reduce torsional vibrations [165]. WEBER developed and tested a magneto-rheological damper that tuned its absorption frequency to the oscillating structure [178]. Another example of active nonlinear devices was presented by MA et al. [114]. They used a fuzzy neural network to control a nonlinear hydraulic adjustable isolator and reduce vibrations in automotive suspensions. A quarter vehicle test bench was excited with white noise and the proposed active nonlinear isolator achieved a 50% reduction of the oscillation's root mean square.

As shown in the paragraphs above, the design of vibration reduction devices is characterized by the vast ingenuity that researchers have embedded in these systems. Furthermore, the research community has thoroughly studied and asserted the capabilities and limitations of linear systems. Therefore, research into active and nonlinear devices has taken on a significant role in the field of vibration reduction. Another example of nonlinear dampers are friction dampers.

1.2.3 Literature Overview of Friction Dampers Research

The targeted use of dry friction to reduce vibrations results in friction dampers. Although friction takes different forms, e.g. viscous friction, friction dampers refer to devices that utilize dry friction, i.e. the unsmeared sliding contact of two solid bodies. To portray the literature of friction dampers, first, the theoretical investigations of such systems are considered. Second, selected practical applications of passive friction dampers are presented. Lastly, the active consideration of friction dampers is detailed.

The effort required to characterize these dampers is high since they are not only nonlinear but also nonsmooth. Mechanical insight into friction dampers is found as early as the 1930s in the work of DEN HARTOG [34]. He studied a one-degree of freedom system with a dry friction element, a spring, and a sinusoidal excitation. He found a solution to the differential equations by considering half of an oscillation and imposing boundary conditions for a periodic solution. Additionally, he considered the system with viscous damping in combination with dry friction. This last system was later on investigated by SHAW. He expanded DEN HARTOG's work by considering a difference in the static and dynamic friction coefficient and carried out a stability and bifurcation analysis [155]. The work of Ruzicka and Derby gave valuable insights into friction dampers in an isolator configuration, in which friction elements were coupled with elastic elements in series and parallel [147], see Fig. 1.10a. Both devices were able to effectively cut off the resonance peak, provided the breakaway force of the friction element is chosen appropriately. The work into the isolator on the right side of Fig. 1.10a was expanded by Fidlin and Lobos, who used averaging methods to describe these mechanisms. They attributed the vibration reduction effect to the switching between stiffnesses, i.e. eigenfrequencies, that resulted due to the sticking of the friction element [54]. A torsional vibrational damper that consists solely of a friction element to dampen vibration and was considered by Alspaugh [3]. He grouped the steady-state motion into three types and concluded that the friction element acted similarly to a low pass filter. Dry friction between the ground and cantilever beams and in between cantilever beams was investigated theoretically and numerically by Dowell and Schwarz [37]. They focused on the variation of the external excitation. With rising excitation magnitude three cases were found: pure stick oscillations, oscillations with stick-slip regimes, and unbounded oscillations. A similar system where the friction element is located in between the beams was recently considered by Krack et al. [99]. However, they focused on modal interactions and their effects on friction damping. The modal interactions were detrimental to the vibration reduction capabilities of dry friction and should therefore be avoided. Ricciardelli and Vickery studied a vibration absorber in combination with a dry friction element between the masses, see Fig. 1.10b. This device was effective at reducing the resonance peak's amplitude provided the breakaway force was tuned correctly to the system [144]. The examples mentioned above share a common denominator, namely the existence of an optimal breakaway force or friction value. Too little friction results in low damping and leads to unbounded oscillations provided there are no other sources of dissipation. Too much friction results in pure stick oscillations that also lead to low damping and high amplitudes. The optimal friction level lies in between and offers maximal damping for a specific excitation level.

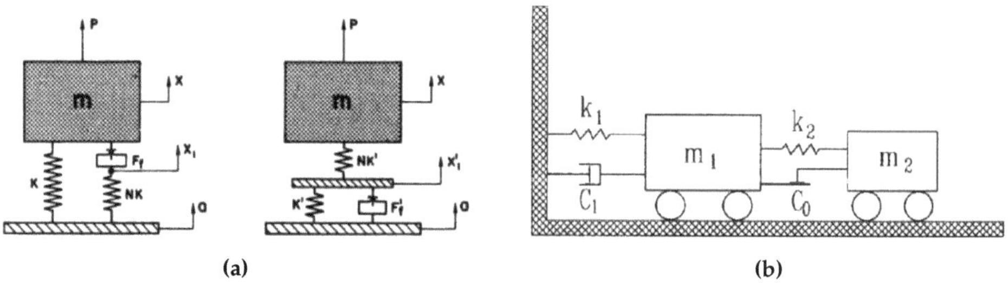

(a) (b)

Figure 1.10: (a) Elastically coupled friction elements in an isolator configuration, source: [147]. (b) A tuned mass damper with a dry friction damping device, source: [144].

A survey on the useful advantages of dry friction in vibration damping and vibration isolation was presented by Ferri [50]. He categorized the applications of friction dampers into four categories: turbomachinery systems, built-up structures, seismic structures, and railroad applications.

Reviews of friction dampers in the context of turbomachinery applications were presented by Popp et al. [140] and more recently by Rizvi et al. [145]. Friction dampers are advantageous in this field since they withstand the extremely harsh conditions in these applications. They are implemented mainly in three ways: between blades and rotor disk, between turbine blades, and in segmented interblade shroud rings [50]. The blade to rotor disk interface usually resembles a fir-tree and is named after it, see Fig. 1.11a. Chan and Tsuba considered a 3D finite element model to describe this region and focused on the effects of the friction coefficient, the clearance, and the normal load [27]. In order to improve the lifetime of turbines, Papanikos and Meguid focused their efforts on the detection and prediction of possible cracks in the blade rotor disk interface and compared simulations with experiments [136]. Meguid et al. studied the design parameters; flank angle, the flank length and the number of teeth, and their effects on the stress in blades and rotor disk [122]. Friction dampers between turbine blades are known as underplatform dampers and are metal elements that are pressed between two blade platforms by the centrifugal force in turbines. These dampers are investigated in two variations: as curved friction dampers and as cottage-roof friction dampers, see Fig. 1.11b. Csaba focused on a curved friction damper with a Winkler foundation and a Jenkin element to model the contact dynamics [30]. A more involved 3D finite element model was proposed by Panning et al. [134] and used to model both curved and cottage-roof dampers. For lower engine orders cottage-roof dampers proved more efficient. Firrone et al. also used a 3D finite element model, however,

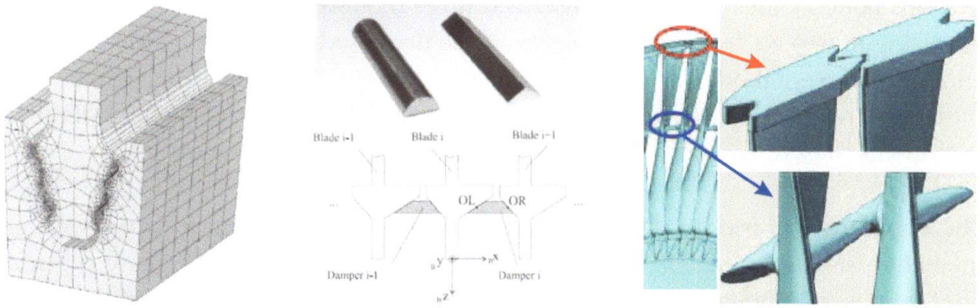

(a) (b) (c)

Figure 1.11: (a) A fir-tree blade rotor disk interface, source: [122]. (b) Curved and wedge friction underplattform dampers, source: [144]. (c) Interblade shroud rings, source: [182]

they considered an asymmetrical underplattform damper. Interblade shroud rings are formed by extrusions along the blade that are in contact with extrusions from adjacent blades and form a ring, see Fig. 1.11c. Bielawa conducted analytic studies into these dampers and concluded that, although there was only dry friction in the contact, these dampers behaved like structural damping for increasing vibrational amplitudes [17]. In the context of shroud ring optimization, Griffin and Labelle used numerical tools to optimize the normal load acting on the contact surfaces [69]. A more recent investigation into these dampers was conducted by Wu et al. and considered 3D finite element models and experiments, to assess the influence of the rotational speed of the turbines of the normal load on the shrouds [182].

Built-up systems represent structures that have been built up by detachable connections, i.e. joints. It is known that friction in such connections provides up to 90% of the damping in structures. However, there is an unavoidable contradiction when optimizing such joints. On the one hand, they should be as tight as possible for a rigid structure, on the other hand, looser joints provide higher structural damping [50]. An overview of the role of friction in mechanical joints was presented by Gaul and Nitsche [66]. It focused on the nonlinear transfer behavior of bolted joint connections. Furthermore, it considered joint models with phenomenological as well as constitutive friction models. The work suggested phenomenological models for systems with reduced degrees of freedom and constitutive friction models for the local description of the joint contact mechanics. A review article by Ibrahim and Pettit [84] focused on the uncertainties involved when modeling bolted joints, e.g. friction, hardness, joint stiffness, and boundary conditions. It presented stochastic and fuzzy finite element models to quantify these uncertainties. Additionally, it considered the identification of bolted joints and the relaxation phenomenon that leads to changes in the joint's characteristics. Bograd et al. [19] gave a more recent overview of the different models which are implemented to describe bolted joints. This work described three models in detail: the node-to-node contact with Jenkins friction elements, finite element models with thin layer elements, and finite element models with zero thickness elements. A common conclusion on all these three reviews is that uncertainties in parameter estimations are a notable limiting factor in the description of joints and their damping in built-up systems.

The application of friction dampers in seismic structures is found in two forms: in moment-resisting frames or in seismic isolation. Examples of both these types of applications are shown in Ferri's review [50]. In his general review of structural control, Houser et al. provided a subchapter dedicated to friction dampers in the context of building applications [81]. A noticeable mention is the Pall device, which was proposed by Pall and Marsch [133]. The device consisted of a friction joint

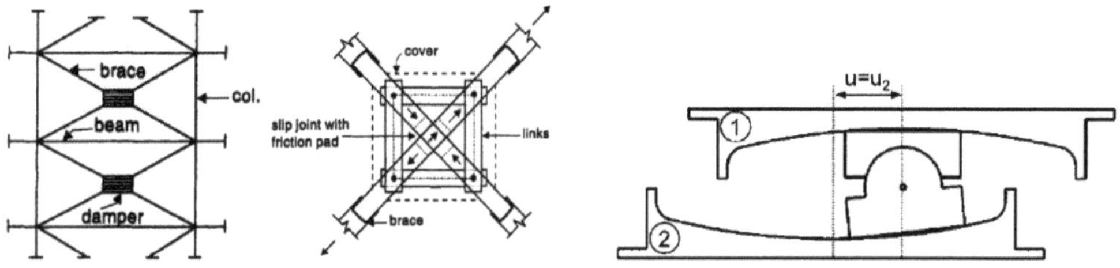

(a) (b)

Figure 1.12: (a) PALL damper in X-brace configuration, source: [133]. (b) Double concave friction pendulum, source: [49].

with slotted holes and was placed at X-bracings of the moment-resisting frame, see Fig. 1.12a. MORGEN and KURAMA also used friction dampers, however, with curved slotted holes to reduce vibrations in beam-to-column joints [127]. In these devices, the breakaway force in the friction joints is set large enough to prevent relative movement during normal service loads and moderate earthquakes. Simultaneously, the preset breakaway force is small enough to allow relative movement during severe seismic excitation. Thus, energy is dissipated and the failure of vital structural components is prevented. These dampers have the disadvantage that they are not self-centering since after large vibrations plastic deformations remain. The problem is addressed by self-centering devices which were initially presented by NIMS et al. [131]. Such mechanisms use friction springs to dissipate energy and return to their original position after excitations subside. Additionally, the have the added feature that the energy dissipated is proportional to the relative displacement. FILIATRAULT et al. considered a similar damper and experimented with a configuration where the damper is placed along a diagonal brace in a moment-resisting frame. The experiments showed the damper was able to dissipate 20% of the energy fed into the structure and thus protected the structure from plastic deformations [56]. KHOO et al. considered such a damper in beam-to-column configuration and showed that residual drift in such arrangements was less than 0.1% [94]. Seismic isolation is achieved in civil engineering structures by allowing a degree of freedom between the buildings and the ground beneath them. A historical review on this subject was published by KELLY [93], whereas a more recent review was given by WARN and RYAN [177]. Initial propositions towards the realization of this isolation combined rubber-bearing isolators and friction elements in a serial arrangement, as considered by CONSTANTINOU TADJBAKHSH [29]. A more modern and widely spread implementation of seismic isolation with friction element is realized in the double concave friction pendulum bearing proposed by FENZ and CONSTANTINOU. It is composed of two concave surfaces, which are separated by a jointed slider, see Fig. 1.12b. In the presented work the concave surfaces were designed in stainless steel and the slider in a PTFE-Composite. Such isolation devices offer a higher number of design parameters, allowing a better performance than their counterparts with a single concave surface [49].

A special variant of friction dampers, the so-called wedge dampers, are used in railroad trucks. Due to their robustness, low cost, and low maintenance they are still investigated today. Furthermore, they play an essential role in reducing vertical and lateral vibrations that lead to derailment. A state-of-the-art review into friction wedge dampers in railroad applications was presented by Wu et al. [183]. Such dampers are composed of a wedge placed between the bolster, which carries the wagon, and the side frame, which is connected to the wheels, see Fig. 1.13. Minimal models with low degrees of freedom were considered by Garnder and Cusumano [64] and by Kaiser et al. [87]. Such minimal models showed the full range of vibration phenomena that occur with dry friction, e.g. stick-slip oscillations, subharmonic bifurcations, and chaos [64]. More involved models with 3D multi-body simulation software were considered by Kovalev et al. [98]. This work compared simulation results with experiments and show an acceptable agreement. Furthermore, the models were able to estimate durability and wear.

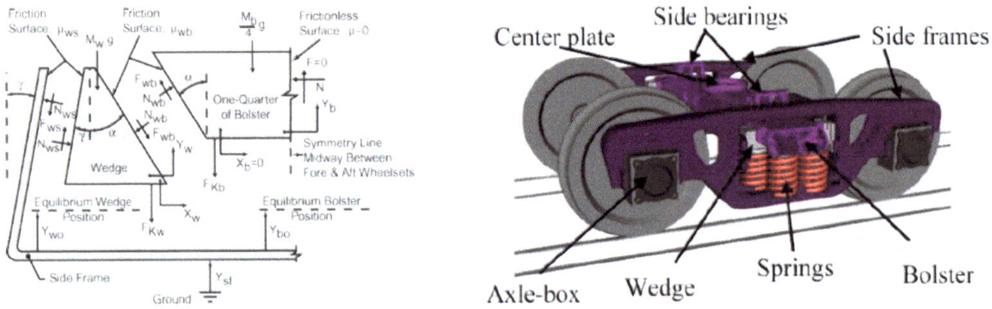

(a) (b)

Figure 1.13: (a) Free body diagram of a friction damper, source: [87]. (b) Model of a Russian 18-100 bogie, source: [98].

Friction-based dampers have also been considered in their semi-active variant in which the normal force is modulated. Lane and Ferri proposed an optimal control and afterward clipped the desired input since the normal force is only positive for unilateral contacts [105]. Dupont et al. used a Ljapunov function based controller to maximize the energy dissipation in the controllable term and derived a corresponding bang-bang control [41]. This controller was very similar to the Skyhook Control proposed by Karnopp, however, it was used for vibration dissipation instead of vibration isolation. Both the clipped optimal control as well as the Skyhook Control were implemented by Gaul et al. to reduce vibrations in the joints of truss structures [65]. For an impulse perturbation, the clipped linear quadratic controller showed a faster reduction of vibration amplitudes. An alternative approach when controlling friction dampers is to prevent sticking since it introduces higher frequencies into the system and prevents dissipation. Such control approaches calculate the necessary normal load for stiction and set the actual normal load slightly under this limit. This strategy was simulated by Lu [111] to reduce seismic vibrations. Experiments with this control strategy also in the context of seismic vibration were performed by Lin et al. [108]. Inaudi proposed an active friction damper with modulated normal force that resulted in a linearly scalable response as with linear viscous damping [85]. To this end, the normal force is chosen proportional to the last local peak. Laffranchi et al. modulated the normal force in the joints of humanoid robots to emulate viscous damping [104].

1.3 Thesis Purpose

As noted above, friction dampers are a reliable and robust damping alternative to conventional tuned mass dampers, even though friction is difficult to describe. This is reflected in the wide range of systems they are implemented in and the amount of literature available on friction dampers. The classic realization of such a vibration reduction device relies on the sliding contact of two nominally flat surfaces, i.e. surfaces that are flat on a macroscopic scale, however, rough on the microscopic scale. In order to expand the capabilities of these systems, the geometric design of the sliding surfaces is considered, e.g. curved underplatform dampers, friction pendulums, and railroad wedge dampers. Further expansions rely on semi-active friction dampers that modulate the normal force between nominally flat sliding surfaces. In general, friction dampers have been mainly used to dissipate energy. However, the effective use of the nonsmooth characteristics of dry friction in the context of vibration reduction has not been thoroughly investigated. There is, therefore, still room for the improvement of such dampers.

The main purpose of this thesis is the design and validation of dry friction based damping mechanisms that rely not solely on dissipation, but instead consider vibration absorption and targeted stick-slip transitions to reduce vibrations. To this end, this work focuses on three secondary objectives.

First, the design of passive devices is considered. Prestressed piecewise defined contact surfaces in combination with a vibration absorber are considered to bring damping, absorption, and targeted stick-slip transitions into one damper. Most friction dampers rely on flat surfaces, wedges, or concave surfaces. The piecewise definition of the contact surfaces allows the damper to operate differently under different amplitude regimes. This enhances the vibration reduction ability of the damper. Additionally, apart from [54, 147], the reviewed literature does not take into account the vibration reduction potential of a targeted stick-slip transition. Since this transition changes the system's eigenfrequencies, it can be used to avoid resonances altogether.

Second, the results from the theoretical investigation and design of such dampers is validated with experiments. A test bench is designed, built, and the dampers are tested. A focus is set on the verification of the qualitative behavior of the deigned dampers.

Third, a further extension of the passive systems is sought after in active form. Control strategies that do not solely focus on damping are designed and applied in combination with the designed passive device structure. The strategies are tested in different scenarios, where transient as well as stationary simulations are considered. Additionally, these strategies are compared to traditional damping focused control.

Focusing on these three secondary objectives contributes to the understanding of dry friction dampers and the fulfillment of the thesis' main purpose.

1.4 Thesis Structure

This work contains 7 chapters and is structured as follows:

Chapter 2 introduces the fundamental concepts required for the understanding of this work. The first section considers periodic orbits and their calculation methods. The second section focuses on the analytical analysis of nonlinear systems with two degrees of freedom. The analysis uses the method of averaging to simplify the system's equations. Lastly, the basic concepts for active control strategies are presented that are used later on.

Chapter 3 considers the wedge damper. A secondary mass is placed between two wedges, that are connected via springs, to reduce vibrations. This arrangement yields a force proportional to the relative displacement. It presents a first step before considering piecewise-defined contact surfaces. To analyze the damper, first, the equations of motion of the damper are derived and a general insight is obtained. Subsequently, both numerical and analytical investigations are carried out. The chapter is concluded with a damper assessment.

Chapter 4 considers the tuned wedge damper. This device represents an extension of the wedge damper. Instead of using wedges, the device utilizes a piecewise-defined contact surface, with a flat segment in the middle, two outer angled segments, and two circular connections. Additionally, as with conventional tuned mass dampers, a spring connects the secondary mass directly to the main system. This arrangement yields different responses at different amplitude levels. As with the wedge damper, first, the equations of motion are derived and a general insight is obtained. In addition, numerical and analytical investigations deepen the understanding of the system. Subsequently, an optimization of the damper parameters is carried out. Finally, an assessment of the device is presented.

Chapter 5 handles experimental investigations into the aforementioned dampers. Firstly, the test bench, as well as the sensor signal post-processing, are described. Afterward, the experimental results are discussed for each of the dampers.

Chapter 6 presents three active extensions of the tuned wedge damper. The control strategies are considered separately in different scenarios and are afterward compared. The considered strategies are the Skyhook Control Strategy, an adaptive multiple model control, and a slow frequency-based control.

Chapter 7 summarizes the work and presents the main conclusions. Additionally, a brief insight into future work is given.

2 Fundamentals

This work analyses two friction dampers with numerical and analytical methods. Additionally, active dampers are considered. The necessary concepts to address these three topics are presented in this chapter. First, the numeric methods are considered for the targeted calculation of periodic solutions. This offers a basis of the numerical calculations in chapters 3 and 4. Second, the averaging method for two degrees of freedom systems is presented. This method is the foundation of the analytical investigations in chapters 3 and 4. Third, the basic concepts of the control strategies, which are the basis of chapter 6, are described.

2.1 Periodic Orbits

Dynamic systems are expressed via a differential equation of the form

$$\dot{\mathbf{x}} = \mathbf{f}(\mathbf{x})\,, \tag{2.1}$$

$$\mathbf{x}(t_0) = \mathbf{x}_0\,, \tag{2.2}$$

where $\mathbf{x} \in \mathbb{R}^n$ is the system's state vector of dimension n, the dot denotes the total derivative with respect to the time t, $\mathbf{f} : \mathbb{R}^n \longrightarrow \mathbb{R}^n$ is a nonlinear function which describes the system's dynamics, and $\mathbf{x}_0$ is the initial condition's vector at t_0. All possible values of the states make up the phase space of a system. Consequently, every momentary state is expressed as a point in the n-dimensional phase space. As time passes, this point moves throughout the phase space and the set of adjacent points is referred to as a trajectory. In this work, the function $\mathbf{f}(\mathbf{x})$ is a nonlinear, autonomous, piecewise continuous function. It is nonlinear because the function can not be expressed as a linear combination of the states. Since the time t is not found explicitly in the equations, the function $\mathbf{f}(\mathbf{x})$ is autonomous. This last characteristic is not restrictive since non-autonomous systems can always be converted to autonomous systems [137]. Furthermore, the system is deterministic if from the initial conditions all the following states of the system are predictable. For arbitrary initial conditions, solving the differential equations yields the transient behavior of the system. As $t \rightarrow \infty$, the behavior of the system is either unbounded and tends to infinity or it approaches a steady-state solution.

2.1.1 Steady-State Solutions of Dynamical Systems

Depending on the characteristics of the steady-state solution, it is categorized into one of four categories [137]: equilibrium point, periodic solution, quasi-periodic solution, and chaotic solution. Since the periodic solution is considered in detail in the following subsection, only the remaining steady-state solutions are briefly described.

Equilibrium points are characterized by $\dot{\mathbf{x}} = \mathbf{0}$. Therefore, the solution stays at the equilibrium point for all time, provided there aren't any external perturbations. Linear systems contain one single equilibrium point, whereas nonlinear systems can have multiple coexisting equilibrium points. Furthermore, except for some special cases, non-autonomous systems have no equilibrium points [137].

When two occurring frequencies in a system have an irrational ratio, quasi-periodic motions occur. Such solutions are separated into periodic functions that interact nonlinearly with each other [137]. Since there is no dominant frequency, the solution does not repeat itself after a given time. Instead, for $t \rightarrow \infty$ the trajectories in the phase space fill up a torus hypersurface in the n-dimensional phase space.

Parker and Chua defined chaos as "a bounded steady-state behavior that is not an equilibrium point, not periodic, and not quasi-periodic" [137]. This definition by negation shows the difficulty regarding the description of chaos. It is characterized by various traits, nevertheless, there is no clear definition for it. Chaos, for example, shows extreme sensitivity to the initial conditions. Therefore, solutions that start close to each other diverge from one another while staying bounded. Furthermore, a possible behavior of chaotic systems is the random alternation between regimes. The explanation of this random characteristic in a deterministic system is still an active research topic.

2.1.2 Periodic Solutions

Periodic solutions repeat themselves after a given time. This is expressed mathematically by the equation

$$\mathbf{x}(t) = \mathbf{x}(t + T) \qquad \forall t > 0 \tag{2.3}$$

where T is the period and is the smallest positive constant, for which Eq. (2.3) holds. Each period has a corresponding fundamental frequency $f = 1/T$ and fundamental angular frequency $\Omega = 2\pi/T$. For autonomous systems, the oscillation frequency is determined by the system characteristics. In monoharmonically excited systems it is determined by some multiple of the excitation frequency. In comparison, the fundamental frequency of multiharmonic excited system is given by a combination of the excitation frequencies.

Portraying the periodic motion $x(t)$ in the phase space yields a closed trajectory. If there exist no other periodic solutions in the vicinity of this trajectory, the solution is considered a limit cycle. Depending on the behavior of transient solutions in the vicinity of the limit cycle, it is considered stable or unstable. If all solutions in a vicinity tend towards the limit cycle, it is stable, otherwise, it is unstable. A detailed description of the stability assessment of periodic motions is presented in [5]. The limit cycle together with its basin of attraction build a subset of the phase space.

2.1.3 Calculation Methods for Periodic Solutions

The most commonly used methods for the targeted calculation of periodic solutions are forward time simulations, shooting methods, collocation methods, and harmonic balance methods. Forward time simulations calculate the dynamic system's behavior until the system reaches the limit cycle. However, this approach is only viable for stable limit cycles. Additionally, this approach is time-consuming, when systems are lightly damped. Shooting methods convert the initial value problem into a boundary value problem by enforcing the periodicity condition. The method starts with an approximated guess for the initial values as well as for the oscillation period. The guess is improved with a NEWTON-RHAPSON Method until the periodicity condition is fulfilled. Collocation methods do not have a single support point, instead, they consider various time points along the trajectory. These time points define intervals and the solutions between the intervals are approximated by polynomials. The method modifies the polynomials to fulfill the continuity and differentiability conditions at the time points, as well as the underlying differential equations of the system. A detailed analysis of shooting and collocations methods is presented by MARX and VOGT in [120]. Lastly, harmonic balance methods approximate the limit cycle via a truncated FOURIER series and consider the problem in the frequency domain. The method minimizes a residuum to calculate the amplitudes of the harmonic base functions. Harmonic balance methods and their application for nonlinear systems are considered by KRACK and GROSS [100]. The following overview of the harmonic balance method is largely based on their work.

Since the harmonic balance method is based on the FOURIER series, the relevant fundamentals are introduced. The periodic vector function $\mathbf{f}(t) \in \mathbb{R}^m$ is expressed as a linear combination of infinite harmonic base functions. An approximation is derived by limiting this infinite series with the truncation order H in the form

$$\mathbf{f}(t) \approx \tilde{\mathbf{f}}(t) = \hat{\mathbf{f}}_{\mathrm{c},0} + \sum_{i=1}^{H} \hat{\mathbf{f}}_{\mathrm{c},i} \cos(i\Omega t) + \hat{\mathbf{f}}_{\mathrm{s},i} \sin(i\Omega t)\,. \tag{2.4}$$

The amplitudes $\hat{\mathbf{f}}_{\mathrm{c},0}$, $\hat{\mathbf{f}}_{\mathrm{c},i}$, and $\hat{\mathbf{f}}_{\mathrm{s},i}$ of the harmonic functions are calculated by evaluating

$$\begin{aligned} \hat{\mathbf{f}}_{\mathrm{c},0} &= \frac{1}{T}\int_0^T \mathbf{f}(t)\mathrm{d}t\,, \\ \hat{\mathbf{f}}_{\mathrm{c},i} &= \frac{2}{T}\int_0^T \mathbf{f}(t)\cos(i\Omega t)\mathrm{d}t\,, \quad \text{and} \quad \hat{\mathbf{f}}_{\mathrm{s},i} = \frac{2}{T}\int_0^T \mathbf{f}(t)\sin(i\Omega t)\mathrm{d}t\,. \end{aligned} \tag{2.5}$$

The transformation between the approximated vector function $\tilde{\mathbf{f}}(t)$ and the amplitudes of the sine-cosine representation $\hat{\mathbf{f}}_{\mathrm{sc},H}$ is written in a compact form as a matrix multiplication

$$\tilde{\mathbf{f}}(t) = \mathbf{h}_{\mathrm{sc},H}(\Omega t)\hat{\mathbf{f}}_{\mathrm{sc},H}\,. \tag{2.6}$$

An advantage of the Fourier series approach is the calculation of derivatives, which are also expressed as a matrix multiplication of the form

$$\dot{\tilde{\mathbf{f}}}(t) = \Omega\boldsymbol{\nabla}\mathbf{h}_{\mathrm{sc},H}(\Omega t)\hat{\mathbf{f}}_{\mathrm{sc},H}\,, \tag{2.7}$$

$$\ddot{\tilde{\mathbf{f}}}(t) = \Omega^2\boldsymbol{\nabla}^2\mathbf{h}_{\mathrm{sc},H}(\Omega t)\hat{\mathbf{f}}_{\mathrm{sc},H}\,. \tag{2.8}$$

In Eqs. (2.6)–(2.8) the amplitudes of the sine-cosine representation $\hat{\mathbf{f}}_{\mathrm{sc},H}$, the matrix of the harmonic base functions $\mathbf{h}_{\mathrm{sc},H}$, and differentiation matrix $\boldsymbol{\nabla}$ are given by

$$\hat{\mathbf{f}}_{\mathrm{sc},H} = \left[\hat{\mathbf{f}}_{\mathrm{c},i}, \hat{\mathbf{f}}_{\mathrm{c},1}, \hat{\mathbf{f}}_{\mathrm{s},1}, \cdots, \hat{\mathbf{f}}_{\mathrm{c},H}, \hat{\mathbf{f}}_{\mathrm{s},H}\right]^{\mathsf{T}}, \tag{2.9}$$

$$\mathbf{h}_{\mathrm{sc},H} = [1\ \cos(\Omega t)\ \sin(\Omega t)\ \cdots\ \cos(H\Omega t)\ \sin(H\Omega t)] \otimes \mathbf{I}_m\,, \tag{2.10}$$

$$\boldsymbol{\nabla} = \mathrm{diag}(0, \boldsymbol{\nabla}_1, \boldsymbol{\nabla}_2, \cdots, \boldsymbol{\nabla}_H) \quad \text{with} \quad \boldsymbol{\nabla}_i = \begin{bmatrix} 0 & i \\ -i & 0 \end{bmatrix}. \tag{2.11}$$

For numerical applications, the discrete Fourier series is required. An effective calculation of the series is implemented with the Fast Fourier Transformation [100].

The starting point of the harmonic balance method is the vector differential equation

$$\mathbf{M}\ddot{\mathbf{q}} + \mathbf{D}\dot{\mathbf{q}} + \mathbf{K}\mathbf{q} + \mathbf{f}_{\mathrm{nl}}(\mathbf{q}, \dot{\mathbf{q}}, t) = \mathbf{f}_{\mathrm{ext}}(\Omega)\,. \tag{2.12}$$

In Eq. (2.12) $\mathbf{M}$, $\mathbf{D}$, and $\mathbf{K}$ represent the mass, damping, and stiffness matrices. With the generalized coordinate vector $\mathbf{q}$ and its derivatives, they make up the linear terms of the system. The term $\mathbf{f}_{\text{nl}}(\mathbf{q}, \dot{\mathbf{q}}, t)$ introduces the influence of the nonlinear terms, whereas $\mathbf{f}_{\text{ext}}(\Omega)$ considers the external excitation. It is assumed that the excitation terms oscillate with the frequency Ω and contain a single harmonic term. The generalized coordinates are approximated by $\mathbf{q} \approx \mathbf{q}_H(t, \hat{\mathbf{q}}_{\text{sc},H}) = \mathbf{h}_{\text{sc},H}(t)\hat{\mathbf{q}}_{\text{sc},H}$ and the residuum is given by

$$\mathbf{r}(t, \ddot{\mathbf{q}}, \dot{\mathbf{q}}, \mathbf{q}) = \mathbf{M}\ddot{\mathbf{q}} + \mathbf{D}\dot{\mathbf{q}} + \mathbf{K}\mathbf{q} + \mathbf{f}_{\text{nl}}(\mathbf{q}, \dot{\mathbf{q}}, t) - \mathbf{f}_{\text{ext}}(\Omega) \stackrel{!}{=} 0\,. \tag{2.13}$$

The basic idea of the harmonic balance method is to calculate the amplitudes $\hat{\mathbf{q}}_{\text{sc},H}$ so that the amplitudes of the residuum until the truncation order H vanish. To this end, the amplitudes of the residuum are calculated as

$$\hat{\mathbf{r}} = \frac{1}{T}\int_0^T \mathbf{h}_{\text{sc},H}(t)\,\mathbf{r}(\ddot{\mathbf{q}}, \dot{\mathbf{q}}, \mathbf{q}, t)\mathrm{d}t = \hat{\mathbf{f}}_{\text{lin}} + \hat{\mathbf{f}}_{\text{nl}} - \hat{\mathbf{f}}_{\text{ext}}\,. \tag{2.14}$$

In Eq. (2.14) the calculation of the residuum amplitudes is broken down into the calculation of the linear terms, the nonlinear terms, and the excitation terms. In the case of a single harmonic excitation, the term $\hat{\mathbf{f}}_{\text{ext}}$ has a single entry in the corresponding row depending on the chosen harmonic function. The linear terms are calculated with

$$\hat{\mathbf{f}}_{\text{lin}} = (\nabla^2 \otimes \Omega^2\mathbf{M} + \nabla^1 \otimes \Omega\mathbf{D} + \nabla^0 \otimes \mathbf{K})\,\hat{\mathbf{q}}_{\text{sc},H}\,. \tag{2.15}$$

The main challenge lies in the calculation of the amplitudes of the nonlinear terms. This challenge is addressed by implementing the Alternating Frequency-Time Scheme [100], depicted in Fig. 2.1. First, this algorithm takes the momentary estimate for $\hat{\mathbf{q}}_{\text{sc},H}$ and calculates the corresponding approximated time series $\tilde{\mathbf{q}}(t)$. Second, this time

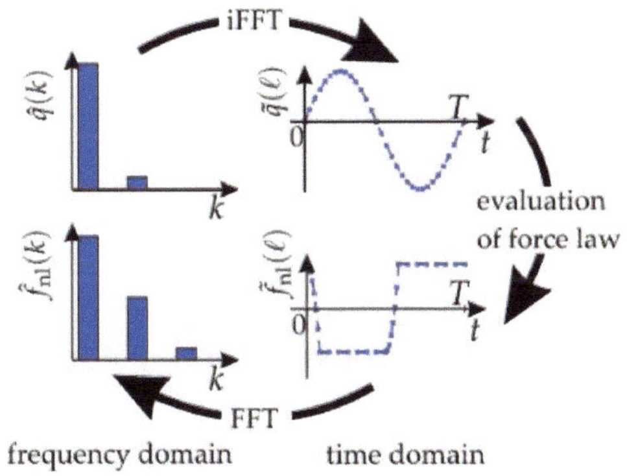

Figure 2.1: Visual representation of the Alternating Frequency-Time scheme, source: [100].

series is used to calculate the corresponding approximated nonlinear forces $\tilde{\mathbf{f}}_{\mathrm{nl}}(t)$ in the time domain. Third, the algorithm calculates the FOURIER series of the approximated nonlinear forces, which yields the amplitudes in $\hat{\mathbf{f}}_{\mathrm{nl}}$.

In order to handle bends and multiple solutions for a single frequency, a continuation method is implemented. To this end, the fundamental frequency Ω is considered a variable, and a new vector $\mathbf{y} = [\hat{\mathbf{q}}_{\mathrm{sc},H}, \Omega]^{\mathsf{T}}$ describes the unknown variables. Consequently, an additional equation is required. According to the pseudo-arc length parametrization, an equation is introduced so that the difference between the solution and the initial guess lies within a hypersphere of radius Δs. Given an initial guess for the variables $\mathbf{y}_0$, the hypersphere requirement is mathematically expressed as

$$(\mathbf{y} - \mathbf{y}_0)^{\mathsf{T}}(\mathbf{y} - \mathbf{y}_0) - \Delta s^2 = 0\,. \tag{2.16}$$

Together with Eq. (2.13), Eq. (2.16) builds the zero point problem

$$\mathbf{R}(\mathbf{y}) = \begin{bmatrix} \hat{\mathbf{f}}_{\mathrm{lin}} + \hat{\mathbf{f}}_{\mathrm{nl}} - \hat{\mathbf{f}}_{\mathrm{ext}} \\ (\mathbf{y} - \mathbf{y}_0)^{\mathsf{T}}(\mathbf{y} - \mathbf{y}_0) - \Delta s^2 \end{bmatrix} = \mathbf{0} \tag{2.17}$$

which is solved by the numeric optimization algorithms. In order to ease the calculations, the algorithm is further expanded with a prediction step. The predictor estimate $\mathbf{y}_{\mathrm{pre}}$ of the next solution is calculated with

$$\mathbf{y}_{\mathrm{pre}} = \mathbf{y}_0 + \Delta s \mathbf{y}_1^*\,, \tag{2.18}$$

$$\text{with} \quad \mathbf{y}_1^* = \frac{\mathbf{y}_1}{\|\mathbf{y}_1\|} \quad \text{and} \quad \left.\frac{\partial \mathbf{R}(\mathbf{y})}{\partial \mathbf{y}}\right|_{\mathbf{y}_0} \mathbf{y}_1 = \mathbf{0}\,. \tag{2.19}$$

In Eq. (2.19) $\mathbf{y}_1$ is calculated with the help of a QR-decomposition. The estimate $\mathbf{y}_{\mathrm{pre}}$ is then used as an initial guess for the next iteration.

The algorithm above is used together with a shooting method for verification purposes. Together they make up a solid framework for the calculation of periodic solutions in nonlinear systems. For more details on the implementation of this algorithm, the reader is referred to [100] and [120].

2.2 Averaging for Two Degrees of Freedom Systems

Although numerical simulations are a basic initial step, analytical investigations offer a deeper insight into the system dynamics. Relationships between the system's behavior and parameters are derived via the averaging method. Developed by KRYLOV and

BOGOULIBOV [102], the basic principle of this method is to decompose the system behavior into their slow and fast oscillating parts. Furthermore, the variables of the slow dynamics are considered constant over one period from the point of view of the fast dynamics. This motivates the averaging of the slow dynamics over the fast oscillating variables. A reduced order system and often simpler equations of motion are obtained. For further details into the averaging method, the reader is referred to the works of MITROPOSLKII [124], SANDERS and VERHULST [150], and FIDLIN [51]. The general approach involving the averaging method for systems with two degrees of freedom is considered in this chapter. To this end, the methods are applied to a general nonlinear dissipative device, see Fig. 2.2.

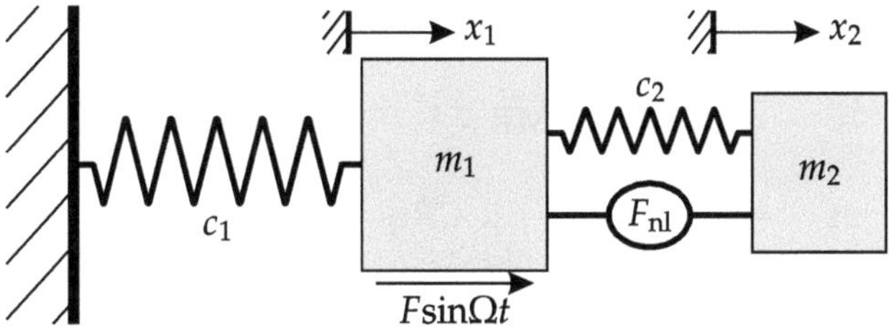

Figure 2.2: A general nonlinear dissipative device.

The exemplary system is composed of the main system and the dissipative device. The main system is represented by the main spring c_1 and the main mass m_1. This main system is excited by the harmonic force with amplitude F and angular frequency Ω. In order to improve its dynamic behavior, the nonlinear dissipative device is attached to the main system. The dissipative device is composed of an auxiliary spring c_2, the nonlinear dissipative element generating the force F_{nl}, and the auxiliary mass m_2. The equations of motion of the system are given by

$$m_1\ddot{x}_1 + c_1x_1 - c_2(x_2 - x_1) - F_{\mathrm{nl}}(x_1, x_2, \dot{x}_1, \dot{x}_2) = F\sin\Omega t\,, \tag{2.20}$$

$$m_2\ddot{x}_2 + c_2(x_2 - x_1) + F_{\mathrm{nl}}(x_1, x_2, \dot{x}_1, \dot{x}_2) = 0\,. \tag{2.21}$$

In order to apply the averaging method, equations Eqs. (2.20) and (2.21) are considered in the dimensionless time τ. The following transformations are applied

$$\begin{gathered}\tau = \omega_{01}t\,,\ \frac{\mathrm{d}(\ )}{\mathrm{d}\tau} = (\ )'\,,\ \omega_{01}^2 = \frac{c_1}{m_1}\,,\ \omega_{02}^2 = \frac{c_2}{m_2}\,,\ \eta = \frac{\Omega}{\omega_{01}}\,,\ \gamma = \frac{m_2}{m_1}\,,\\ p^2 = \frac{c_2}{m_1\omega_{01}^2} = \frac{c_2}{c_1}\,,\ \varepsilon f_{\mathrm{nl}} = \frac{F_{\mathrm{nl}}(x_1, x_2, \dot{x}_1, \dot{x}_2)}{m_1\omega_{01}^2}\,,\ \varepsilon f = \frac{F}{m_1\omega_{01}^2}\,,\ \varepsilon \ll 1\,.\end{gathered} \tag{2.22}$$

As noted above, ε is a small parameter and implies small nonlinear dissipative forces and small excitation amplitudes. Applying the transformations above results in

$$x_1'' + x_1 - p^2(x_2 - x_1) = \varepsilon \left(f \sin\eta\tau + f_{\mathrm{nl}}(x_1, x_2, x_1', x_2')\right) = \varepsilon f_{\mathrm{nl},1}\,, \tag{2.23}$$

$$\gamma x_2'' + p^2(x_2 - x_1) = -\varepsilon \left(f_{\mathrm{nl}}(x_1, x_2, x_1', x_2')\right) = \varepsilon f_{\mathrm{nl},2}\,. \tag{2.24}$$

The next step in the groundwork for the averaging method is the decoupling of the equations. Hence, a modal transformation is applied to Eqs. (2.23) and (2.24), which yields

$$\mathbf{M} = \begin{bmatrix} 1 & 0 \\ 0 & \gamma \end{bmatrix}, \quad \mathbf{C} = \begin{bmatrix} 1+p^2 & -p^2 \\ -p^2 & p^2 \end{bmatrix}, \quad \mathbf{f} = \begin{bmatrix} f_{\mathrm{nl},1} \\ f_{\mathrm{nl},2} \end{bmatrix}, \quad \mathbf{x} = \begin{bmatrix} x_1 \\ x_2 \end{bmatrix}, \tag{2.25}$$

$$\mathbf{R} = \begin{bmatrix} r_{11} & r_{12} \\ r_{21} & r_{22} \end{bmatrix}, \quad \mathbf{q} = \mathbf{R}\mathbf{x}, \quad \mathbf{R}^{\mathsf{T}}\mathbf{M}\mathbf{R} = \mathbf{I}, \quad \mathbf{R}^{\mathsf{T}}\mathbf{C}\mathbf{R} = \mathrm{diag}(\eta_{01}^2, \eta_{02}^2)\,, \tag{2.26}$$

$$\mathbf{M}\mathbf{x}'' + \mathbf{C}\mathbf{x} = \varepsilon\mathbf{f} \quad \longrightarrow \quad \mathbf{R}^{\mathsf{T}}\mathbf{M}\mathbf{R}\mathbf{q}'' + \mathbf{R}^{\mathsf{T}}\mathbf{C}\mathbf{R}\mathbf{q} = \varepsilon\mathbf{R}^{\mathsf{T}}\mathbf{f}\,. \tag{2.27}$$

The vector $\mathbf{q} = [q_1, q_2]^{\mathsf{T}}$ describes the modal coordinates and the modal matrix $\mathbf{R}$ is calculated considering the unperturbed system, i.e. $\varepsilon = 0$. The modal transformation leads to

$$q_i'' + \eta_{0i}^2 q_i = \varepsilon \left(r_{1i} f_{\mathrm{nl},1}(q_i, q_j) + r_{2i} f_{\mathrm{nl},2}(q_i, q_j)\right) = \varepsilon g(q_i, q_j) \quad (i,j) = \{(1,2),(2,1)\}. \tag{2.28}$$

Since the terms on the right-hand side of Eq. (2.28) are of $\mathcal{O}(\varepsilon)$, the system is weakly coupled. A complete modal decoupling proposed by Fidlin and Gafur in [53, 62] is applied. In essence, only the corresponding modal coordinate is considered for each modal equation. However, this assumption is only valid if the eigenfrequencies η_{01} and η_{02} are far enough from each other, i.e. $|\eta_{02} - \eta_{01}| = \mathcal{O}(1)$ [186]. Under this assumption, the unrelated modal coordinate is of the magnitude order ε in the vicinity of the considered eigenfrequency. Therefore, its influence is first observed in the terms of $\mathcal{O}(\varepsilon^2)$ in Eq. (2.28). This is made evident by considering the dependency $q_j(\varepsilon) = \varepsilon\tilde{q}_j$ and a Taylor expansion about the point $\varepsilon = 0$ up to the terms of $\mathcal{O}(\varepsilon)$, which is given by

$$q_i'' + \eta_{0i}^2 q_i = \left(g(q_i, q_j(\varepsilon)) + \varepsilon \frac{\partial g(q_i, q_j(\varepsilon))}{\partial q_j(\varepsilon)} \frac{\partial q_j(\varepsilon)}{\partial \varepsilon} \right)\Bigg|_{\varepsilon=0} (\varepsilon - 0) + \mathcal{O}(\varepsilon^2)\,. \tag{2.29}$$

An evaluation of equation Eq. (2.29) ultimately results in

$$q_i'' + \eta_{0i}^2 q_i \approx \varepsilon \left(r_{1i} f_{\mathrm{nl},1}(q_i, 0) + r_{2i} f_{\mathrm{nl},2}(q_i, 0)\right) \quad i = \{1,2\}\,. \tag{2.30}$$

In order to obtain the system in standard form for averaging, the following steps are taken. First, a VAN DER POL Transformation, introduced as $q_i = A_i \sin(\varphi_i)$ and $q_i' = A_i \eta_{0i} \cos(\varphi_i)$, is applied to the modal equations. Second, the phase differences $\psi_i = \varphi_i - \eta\tau$ are considered as slow-changing variables. Third, the response of the system is investigated in the vicinity of the eigenfrequencies, i.e. $\eta_{0i} - \eta = \varepsilon\delta_i$. These transformations lead to

$$A_i' = \varepsilon \frac{r_{1i} f_{\mathrm{nl},1}(A_i, \psi_i, \varphi_i) + r_{2i} f_{\mathrm{nl},2}(A_i, \psi_i, \varphi_i)}{\eta_{0i}} \cos(\varphi_i)\,, \tag{2.31}$$

$$\psi_i' = \varepsilon \left(\delta_i - \frac{r_{1i} f_{\mathrm{nl},1}(A_i, \psi_i, \varphi_i) + r_{2i} f_{\mathrm{nl},2}(A_i, \psi_i, \varphi_i)}{A_i \eta_{0i}} \sin(\varphi_i) \right), \tag{2.32}$$

$$\varphi_i' = \eta_{01} - \varepsilon \frac{r_{1i} f_{\mathrm{nl},1}(A_i, \psi_i, \varphi_i) + r_{2i} f_{\mathrm{nl},2}(A_i, \psi_i, \varphi_i)}{A_i \eta_{0i}} \sin(\varphi_i)\,. \tag{2.33}$$

Equations (2.31) and (2.32) represent the system in standard form, whereas Eq. (2.33) denotes the fast oscillating variables. The equations in standard form are averaged over the fast rotating phases φ_1 and φ_2 over one oscillation period, i.e. 2π. Additionally, the averaged variables for the amplitude $\bar{A}_i'$ and phase difference $\bar{\psi}_i'$ are introduced. For the exemplary system, the averaged variables are given by

$$\bar{A}_i' = \langle A_i' \rangle_{\varphi_i} = \frac{1}{2\pi} \int_0^{2\pi} A_i' \mathrm{d}\varphi_i\,, \tag{2.34}$$

$$\bar{\psi}_i' = \langle \psi_i' \rangle_{\varphi_i} = \frac{1}{2\pi} \int_0^{2\pi} \psi_i' \mathrm{d}\varphi_i\,. \tag{2.35}$$

The equations above describe the slow system dynamics and strickly speaking the difference between the full and the averaged system remains of $O(\varepsilon)$ for a time $O(1/\varepsilon)$ [102]. However, ECKHAUS [44] proposed theorems that showed an error estimate extension to $t \longrightarrow \infty$ for the periodic case, provided the right-hand side of Eqs. (2.34) and (2.35) are uniformly bounded and LIPSCHITZ-continuous with respect to A_i and ψ_i. The averaged equations are thus considered for the stationary solution ($\bar{A}_i' = 0$ and $\bar{\psi}_i' = 0$) and verified with the numerical solution. Additionally, relationships between the maximum amplitude and the excitation force are found, see sections 3.3 and 4.3.

The last step is to reconstruct the solution in the original coordinates x_1 and x_2. In the case of linear systems, the transformation takes the form

$$A_{x_1} = \sqrt{r_{11}^2 A_1^2 + r_{12} A_2^2 + 2 r_{11} r_{12} A_1 A_2 \left(\cos\theta_1 \cos\theta_2 + \sin\theta_1 \sin\theta_2\right)}\,, \tag{2.36}$$

$$A_{x_2} = \sqrt{r_{21}^2 A_1^2 + r_{22} A_2^2 + 2 r_{21} r_{22} A_1 A_2 \left(\cos\theta_1 \cos\theta_2 + \sin\theta_1 \sin\theta_2\right)}\,. \tag{2.37}$$

However, an explicit solution of the amplitudes A_i is not guaranteed since Eqs. (2.34) and (2.35) are nonlinear. Furthermore, since the calculated solutions are only valid in the vicinity of the corresponding eigenfrequencies, an application outside this frequency range is not valid. Therefore, the reconstruction is frequency-dependent and results in

$$\eta \approx \eta_{01} \quad \rightarrow \quad A_{x_1} \approx |r_{11}| A_1 \,, \quad A_{x_2} \approx |r_{21}| A_1 \,, \tag{2.38}$$

$$\eta \approx \eta_{02} \quad \rightarrow \quad A_{x_1} \approx |r_{12}| A_2 \,, \quad A_{x_2} \approx |r_{22}| A_2 \,. \tag{2.39}$$

The procedure above describes the foundation of the analytical investigations. These equations are analyzed, and essential relations are derived from them. Detailed practical applications of these equations are presented in chapters 3 and 4.

2.3 Control Strategies for Friction Dampers

As mentioned in section 1.2.3 friction dampers have been used in association with semi-active control strategies. In this subchapter, the fundamental concepts of two control strategies are discussed in detail. Firstly, the Skyhook Control Strategy developed by Karnopp et al. [91] is considered. Furthermore, the relationship between this strategy and the work of Dupont et al. [41] is presented. Subsequently, the basic structure of adaptive control with a reference model is discussed. These concepts build the basis of the active damper investigations, which are discussed in chapter 6.

2.3.1 Skyhook Control

The idea of the Skyhook Control Strategy is to emulate a damper attached to an inertial reference point, even though the damper itself is placed between two moving bodies. However, this emulation is not always viable. To avoid exacerbating the oscillations of

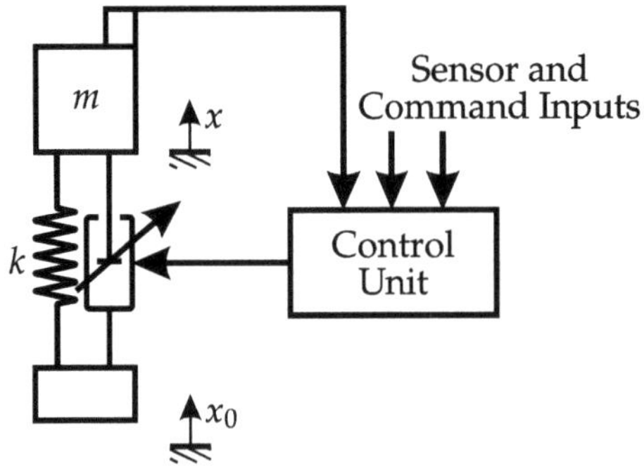

Figure 2.3: A semi-active isolator system, modified from source: [91].

the considered mass, the damper force is set to zero whenever damping is not possible. This leads to a damping-focused strategy. In [91] the damping constant is varied in combination with the system in Fig. 2.3 resulting in the damper force

$$F_d = \begin{cases} d_{\text{Sky}}(\dot{x} - \dot{x}_0), & \dot{x}(\dot{x} - \dot{x}_0) > 0 \\ 0, & \dot{x}(\dot{x} - \dot{x}_0) < 0 \end{cases} = \begin{cases} d_{\text{d}}\dot{x}, & \dot{x}(\dot{x} - \dot{x}_0) > 0 \\ 0, & \dot{x}(\dot{x} - \dot{x}_0) < 0 \end{cases} . \tag{2.40}$$

In Eq. (2.40) d_{Sky} represents the controlled damping coefficient, whereas d_{d} represents the desired damping coefficient of the emulated damper. Alternatively, Dupont et al. motivated this control strategy via Ljapunov functions [41]. A Ljapunov function V based on the energy associated with main mass m in Fig. 2.3 yields

$$V = \frac{1}{2}k(x - x_0)^2 + \frac{1}{2}m\dot{x}^2 > 0\,, \tag{2.41}$$

$$\dot{V} = k(x - x_0)(\dot{x} - \dot{x}_0) + m\dot{x}\ddot{x} = -k(x - x_0)\dot{x}_0 - F_d\dot{x}\,. \tag{2.42}$$

Focusing on the controllable terms in the Ljapunov derivative results in

$$\dot{V}_d = -F_d\dot{x} = -d_{\text{Sky}}(\dot{x} - \dot{x}_0)\dot{x}\,. \tag{2.43}$$

In order to provide damping, the Ljapunov derivative of the controllable terms must be negative. To avoid exacerbating the oscillations of the main mass, it cannot be positive. Since the damping coefficients d_{Sky} and d_{d} are strictly positive, this yields

$$d_{\text{Sky}} = \begin{cases} \dfrac{d_{\text{d}}\dot{x}}{\dot{x} - \dot{x}_0}, & \dot{x}(\dot{x} - \dot{x}_0) > 0 \\ 0, & \dot{x}(\dot{x} - \dot{x}_0) \leq 0 \end{cases} . \tag{2.44}$$

Equation (2.44) leads to large damping constants in the vicinity of $\dot{x} \approx \dot{x}_0$. This is addressed by considering a limit on the input variable, e.g. $d_{\text{Sky}} \in [0, d_{\text{max}}]$. If the maximal possible damping is desired, the control rule is simplified to

$$d_{\text{Sky}} = \begin{cases} d_{\text{max}}, & \dot{x}(\dot{x} - \dot{x}_0) > 0 \\ 0, & \dot{x}(\dot{x} - \dot{x}_0) \leq 0 \end{cases} . \tag{2.45}$$

This approach does not ensure the stability of the whole system, since only a part of the Ljapunov derivative is considered. It does. however, provide dissipation whenever possible and ensures the control strategy does not exacerbate the system's oscillations. These qualities are also ensured in the work of Karnopp et al., although Dupont et al. formulated the strategy in a more general sense.

2.3.2 Adaptive Control

Adaptive control is characterized by the extension of the classical control structure with one feedback loop by an additional adaptation loop. The adaptation loop is tasked with the adjustment of the controller parameters according to an adaption law. This allows the design of a much more flexible controller that can adapt its behavior to changes in the environment. This chapter focuses on two adaptive control methods: gain scheduling and model reference adaptive control.

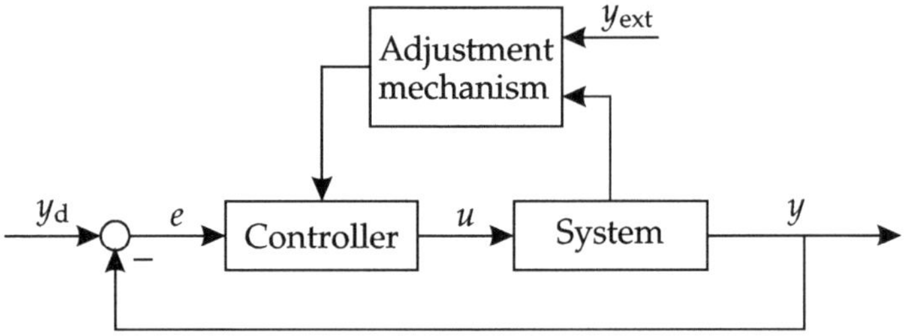

Figure 2.4: Gain scheduling adaptive control, modified from source: [152].

Gain scheduling extends the classical feedback loop with the help of an external measurement, see Fig. 2.4. This measurement provides additional insight into the system's behavior. Thus, the parameter adaption is given as a function of the additional measurement. Since the only limitation lies in the sampling time of the external signal, this method allows a fast change in the controller parameters. However, there is a drawback to this approach. Since the adaptation loop is open, there is no real intelligence or "learning" ability in the system [152]. The gain scheduler requires precise knowledge of the system, and can only adapt if the changes in the environment are accounted for. Therefore, it cannot improve itself against unforeseen changes.

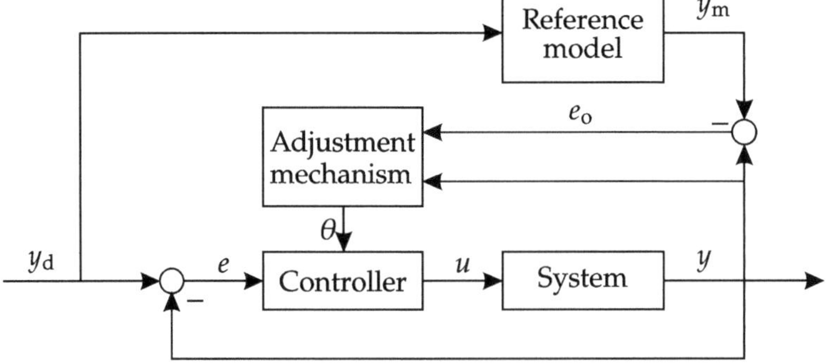

Figure 2.5: Model reference adaptive control, modified from source: [152].

In contrast to gain scheduling, the model reference adaptive control has a closed adaptation loop and with it intelligence within the control structure. This approach relies on a reference model, see Fig. 2.5. The reference model contains the desired behavior of the system. The parameters of the controller are adjusted so that the system asymptotically matches the reference model. To this end, the adjustment mechanism changes the controller parameters θ and is calculated to reduce e_{o}^2. This results in an adaptation law of the form

$$\frac{\partial \theta}{\partial t} = -2 g e_{\mathrm{o}} \frac{\partial y}{\partial \theta} \,. \tag{2.46}$$

As the controller parameters θ tend to the ideal parameters θ^*, the output error e_{o} tends to zero. However, the partial derivatives of the system output depend on the unknown and in some cases varying system parameters. This obstacle is avoided by replacing the unknown system parameters with their identified estimates. Alternatively, approaches via Ljapunov functions are also formulated to ensure the output error and the difference $\theta - \theta^*$ tends to zero. For a detailed description concerning the calculation of the adaptation law, the reader is referred to the work of Sastry and Bodson [152].

3 The Wedge Damper

As previously mentioned, the analysis of the wedge damper, see Fig. 3.1, presents a first step before taking into account dampers with piecewise defined contact surfaces. A general description of the damper is presented in section 3.1. This offers a first impression of the damper's behavior. These initial insights are expanded by numerical simulations in section 3.2. To obtain key relationships between the damper behavior and its parameters, analytical investigations are carried out in section 3.3. Finally, an assessment of the damper is presented in section 3.4. Works into the characteristics of this damper are presented in [184, 185]. The minimal model of this damper is based on a patent from the Schaeffler AG [75, 180].

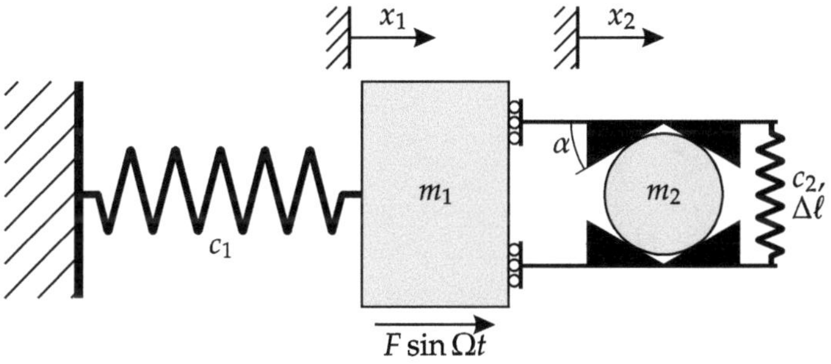

Figure 3.1: The schematic model of the wedge damper.

3.1 General Damper Description

The wedge damper is attached to the main system, represented by the primary mass m_1 and the primary spring c_1. The main system is excited by the harmonic force with amplitude F and angular frequency Ω, and it is the damper's task to improve the system dynamics. The wedge damper is composed of a secondary mass m_2 placed between two wedge-shaped contact surfaces with the wedge angle α. The friction coefficient μ describes the dry friction between the secondary mass and the contact surfaces.

Furthermore, the wedges are pressed onto the secondary mass by the secondary spring c_2 which has a prestress displacement $\Delta\ell$. Due to the damper design, a relative movement between the primary and secondary masses forces the wedges apart from each other. This increases the force in the secondary spring and thus the contact forces between the secondary mass and the wedges. Thus, the contact forces increase with the absolute value of the relative displacement. In contrast, classical alternatives rely solely on dry friction, nominally flat surfaces, and constant contact forces.

Due to the nature of dry friction, the dynamics of this system are described by two sets of equations. When the system slips, there is a relative movement between the masses, the system has two degrees of freedom, and it is described by the differential equations

$$m_1\ddot{x}_1 + c_1 x_1 - F_{\mathrm{WD}} = F \sin \Omega t \,, \tag{3.1}$$

$$m_2\ddot{x}_2 + F_{\mathrm{WD}} = 0 \,. \tag{3.2}$$

Where x_1 and x_2 are the position coordinates of the primary and secondary mass, the dot represents the total derivative with respect to the time t, and F_{WD} is the horizontal wedge damper force acting on both masses. The wedge damper force is obtained with the free body diagram presented in Fig. 3.2.

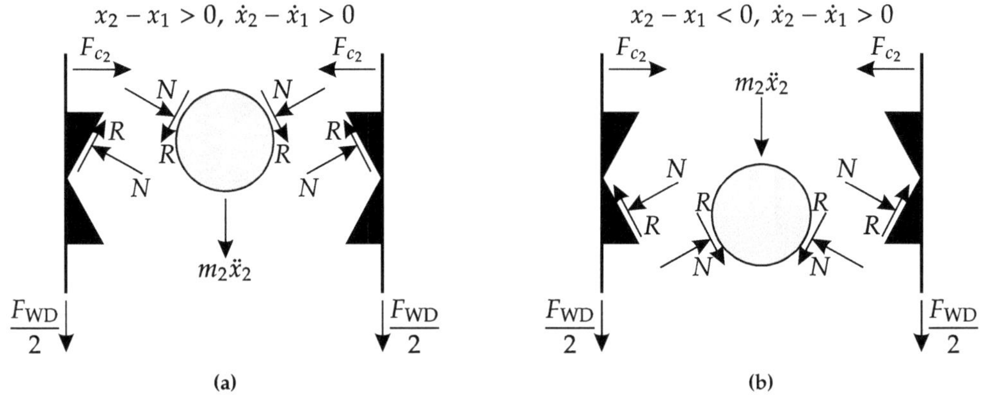

Figure 3.2: Free body diagram of the wedge contact and secondary mass for: (a) $x_2 - x_1 > 0,\ \dot{x}_2 - \dot{x}_1 > 0$ and (b) $x_2 - x_1 < 0,\ \dot{x}_2 - \dot{x}_1 > 0$.

The sums of forces resulting from Figs. 3.2a and 3.2b are summarized in two equations. These are given with the sign-function of the relative displacement and result in

$$R\cos\alpha + N\sin\alpha\,\mathrm{sign}(x_2 - x_1) - \frac{F_{\mathrm{WD}}}{2} = 0 \,, \tag{3.3}$$

$$R\sin\alpha\,\mathrm{sign}(x_2 - x_1) - N\cos\alpha + F_{c_2} = 0 \,. \tag{3.4}$$

Where N and R are the friction forces in the contact and F_{c_2} is the spring force. With the relation for COULOMB friction $R = \mu N \operatorname{sign}(\dot{x}_2 - \dot{x}_1)$ and the secondary spring force $F_{c_2} = c_2(2\tan\alpha|x_2 - x_1| + \Delta\ell)$, the equations above are solved for the normal contact force and the wedge damper force. The latter results in

$$F_{\text{WD}} = 2c_2\left(2\tan\alpha|x_2 - x_1| + \Delta\ell\right)\frac{\tan\alpha\,\operatorname{sgn}(x_2 - x_1) + \mu\operatorname{sgn}(\dot{x}_2 - \dot{x}_1)}{1 - \mu\tan\alpha\,\operatorname{sgn}(x_2 - x_1)\operatorname{sgn}(\dot{x}_2 - \dot{x}_1)}\,. \tag{3.5}$$

In contrast, when the system sticks, there is no relative movement between the masses. The movement of the sticking system are described by one differential equation

$$(m_1 + m_2)\ddot{x}_1 + c_1 x_1 = F\sin\Omega t\,. \tag{3.6}$$

Since both masses move as one, the system has consequently one degree of freedom and the eigenfrequency $\omega_{\text{st}} = \sqrt{c_1/(m_1 + m_2)}$.

Additionally the system dynamics are described by sticking conditions, which expressed as inequalities. Once these conditions are fulfilled the system transitions from the slipping to the sticking state, and vice versa if the conditions are broken. Depending on the system's transition into sticking, two different sticking conditions are evaluated. However, regardless of the transition mechanism the relative velocity $\dot{x}_2 - \dot{x}_1 = 0$ at the transition. The system sticks either at the middle point between the wedges where the slope changes abruptly ($x_2 - x_1 = 0$) or on the surfaces of two wedges ($x_2 - x_1 \neq 0$). The dominant transition is deduced later on after a closer analysis of the damper force.

For the first case, sticking at $x_2 - x_1 = 0$, the wedge damper force in Eqs. (3.1) and (3.2) is replaced with the momentary stiction force H and it is calculated to ensure $\dot{x}_2 - \dot{x}_1 = 0$. The maximal stiction force $H_{\max}$ is given by the limit of F_{WD} as $x_2 - x_1$ and $\dot{x}_2 - \dot{x}_1$ approach 0 from above. This is the maximum stiction force that the damper generates at $x_2 - x_1 = 0$. This procedure results in

$$H = \frac{m_2}{m_1 + m_2}(F\sin\Omega t - c_1 x_1)\,, \tag{3.7}$$

$$H_{\max} = \lim_{\substack{x_2 - x_1 \to 0^+ \\ \dot{x}_2 - \dot{x}_1 \to 0^+}} F_{\text{WD}} = 2c_2\Delta\ell\frac{\tan\alpha + \mu}{1 - \mu\tan\alpha}\,. \tag{3.8}$$

$$|H| =\leq H_{\max}\,, \quad x_1 = x_2\,, \quad \text{and} \quad \dot{x}_1 = \dot{x}_2\,. \tag{3.9}$$

In the second case, sticking at $x_2 - x_1 \neq 0$, the stiction is determined by the resulting contact force of R and N, which is generated by the necessary stiction force and the spring force. The sticking condition is fulfilled, when the resulting friction force lies within the friction cone given by the friction coefficient μ. In this case, the friction force R is not given by the Coulomb friction and is instead a constraint force. Additionally, F_{WD} is also replaced by H. Furthermore, Eqs. (3.3) and (3.4) are solved for R and N and the sticking conditions result in

$$R = F_{c_2} \sin\alpha - \frac{H}{2}\cos\alpha\,, \tag{3.10}$$

$$R_{max} = \mu N = \mu\left(F_{c_2}\cos\alpha + \frac{H}{2}\sin\alpha\right), \tag{3.11}$$

$$|R| < R_{max}\,, \quad \text{and} \quad \dot{x}_1 = \dot{x}_2\,. \tag{3.12}$$

Preliminary characteristics of the system are derived by evaluating the damper force while sliding. Due to the sign-function, the force F_{WD} is nonsmooth both on a displacement and on a velocity level. Thus, the force is highly nonlinear. The force determines the damper's behavior during the slip phase and is separated into a conservative part $F_{WD,c}$ and a dissipative part $F_{WD,d}$. These forces are given by

$$F_{WD,c} = F_{WD}(\mu = 0) = 2\tan\alpha c_2\left(2\tan\alpha(x_2 - x_1) + \Delta\ell\,\mathrm{sgn}(x_2 - x_1)\right), \tag{3.13}$$

$$\begin{aligned} F_{WD,d} &= F_{WD} - F_{WD,c} \\ &= 2c_2\left(2\tan\alpha|x_2 - x_1| + \Delta\ell\right)\frac{\mu(1+\tan^2\alpha)\mathrm{sgn}(\dot{x}_2 - \dot{x}_1)}{1 - \mu\tan\alpha\,\mathrm{sgn}(x_2 - x_1)\mathrm{sgn}(\dot{x}_2 - \dot{x}_1)}\,. \end{aligned} \tag{3.14}$$

The conservative part of F_{WD} is composed of a spring force proportional to the relative displacement between the masses and a direction-dependent prestress. The effective secondary stiffness of the spring force is $c_{2,eff} = 4c_2\tan^2\alpha$. The dissipative terms are

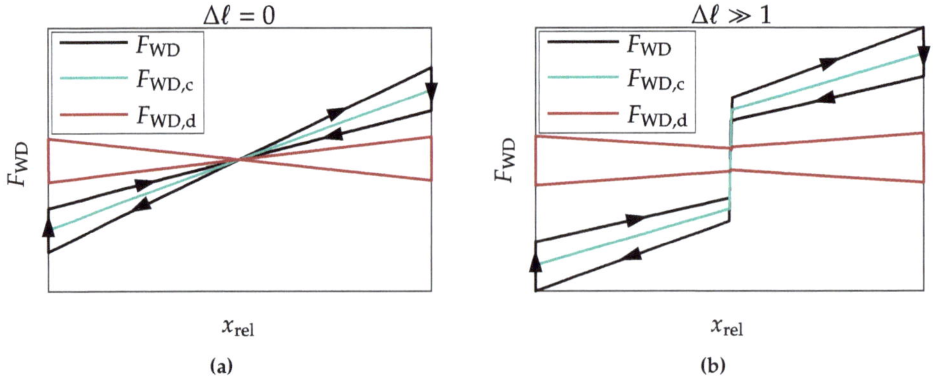

Figure 3.3: Exemplary wedge damper force for (a) vanishing prestress displacement and (b) large prestress displacement.

proportional to the amplitude of the relative displacement between the masses, i.e. $F_{\mathrm{WD,d}} \sim A_{\mathrm{rel}}$. Two qualitative curves of the damper forces are plotted against the relative displacement $x_{\mathrm{rel}} = x_2 - x_1$ in Fig. 3.3. For a vanishing prestress there is no discontinuity at $x_{\mathrm{rel}} = 0$ and no sticking at this position. In contrast, a large prestress results in a proportional discontinuity at $x_{\mathrm{rel}} = 0$ allowing sticking at this point. Discontinuities are also observed at $x_{\mathrm{rel}} \neq 0$ due to the change in the sign of the relative velocity, c.f. Eq. (3.5). However, these discontinuities are not as pronounced as the one in the middle of the hysteresis. This indicates that sticking at $x_{\mathrm{rel}} = 0$ is the dominant transition mechanism between sliding and sticking for the investigated parameters.

An approximation for the dissipated work of the damper $W_{F_{\mathrm{WD,d}}}$ is obtained by assuming a harmonic oscillation with a phase $\varphi = \Omega t$ for the relative coordinate $x_{\mathrm{rel}} \approx A_{\mathrm{rel}} \sin\varphi$ and the relative velocity $\dot{x}_{\mathrm{rel}} \approx \Omega A_{\mathrm{rel}} \cos\varphi$. An evaluation of the work integral yields

$$W_{F_{\mathrm{WD,d}}} \approx -\int F_{\mathrm{WD,d}}(x_{\mathrm{rel}}, \dot{x}_{\mathrm{rel}})\, \mathrm{d}x_{\mathrm{rel}} = -8c_2\mu(\tan\alpha A_{\mathrm{rel}}^2 + \Delta\ell A_{\mathrm{rel}}) \frac{1+\tan^2\alpha}{1-\mu^2\tan^2\alpha}\,. \quad (3.15)$$

The dissipated work is proportional to the square value of the relative displacement amplitude, i.e. $W_{F_{\mathrm{WD,d}}} \sim A_{\mathrm{rel}}^2$. This is also the case with viscous damping, and a similar behavior is expected. Therefore, this vibration reduction mechanism is considered a pseudo-viscous damper. Compared to the dissipated energy of a viscous damper ($W_{\mathrm{VD}} = \pi d \Omega A_{\mathrm{rel}}^2$), the dissipated energy in the wedge damper is independent of the excitation frequency. This is especially advantageous at low frequencies.

Before presenting the numerical and analytical investigations, it is helpful to identify the stick and slip ranges. While sticking, the system is completely linear and its solution is known. With this solution, an estimate for the adherence to the stiction condition is obtained for the dominant stick-slip transition. A solution of the form $x_1 = A\sin\varphi$ is plugged into the stiction inequality Eq. (3.9) and a limit amplitude is estimated. The estimate is calculated with

$$\begin{aligned} |H| &= \frac{m_2}{m_1+m_2}|F\sin(\Omega t) - c_1 A\sin\varphi| = m_2|\ddot{x}_1| = m_2\Omega^2 A|\sin\varphi|, \\ &\leq m_2\Omega^2 A \leq H_{\max}, \\ A &\leq \frac{H_{\max}}{m_2\Omega^2} = \frac{2c_2\Delta\ell(\tan\alpha+\mu)}{m_2\Omega^2(1-\mu\tan\alpha)} = A_{\mathrm{L}} \end{aligned} \quad (3.16)$$

where A_{L} is the limit amplitude of the linear system. Equation (3.16) describes the limit amplitude for sticking at $x_2 - x_1 = 0$ and offers an acceptable approximation of the transition between the linear and nonlinear ranges.

3.2 Numerical Investigations

The numerical investigations expand the initial insights into the behavior of the wedge damper. This section is divided into two parts. First, parameter variations are carried out. The algorithm described in section 2.1 is implemented to calculate the frequency response function of the system. The variation of the design parameters leads to initial insight into the parameters' effect on the system. In the second part, a stability analysis of selected frequency response functions is presented. A focus is set on chosen unstable solutions and POINCARÉ maps are constructed to characterize the steady-state solutions. Unless specified otherwise, the standard parameters for the investigations are:

$$m_1 = 1\,\mathrm{kg},\ m_2 = 0.1\,\mathrm{kg},\ c_1 = 1\,\mathrm{N/m},\ c_2 = 0.1\,\mathrm{N/m},\ \alpha = 30^\circ,$$
$$\Delta\ell = 0.01\,\mathrm{m},\ \mu = 0.01,\ F = 0.01\,\mathrm{N}.$$

The parameter studies of the variations of the secondary spring c_2 and the wedge angle α are presented respectively in Figs. 3.4 and 3.5. By comparing these two studies, similarities in their effect are found. Both parameters have a similar effect as the variation of the auxiliary spring in a standard vibration absorber. An increase in the stiffness or the angle increases the first resonance peak while shifting it towards higher frequencies. The inverse effect is observed at the second resonance peak. The reason for this lies in the change in the effective secondary stiffness these parameters cause. These parameters optimize the performance at the resonance peaks or determine the absorption frequency. As with linear systems, simultaneous optimization of both cannot be achieved.

The variation of the friction coefficient μ is presented in Fig. 3.6. This parameter influences the damping forces $F_{\mathrm{WD,d}}$ as well as the breakaway force $H_{\max}$ of the damper. An increase in the friction coefficient raises the damping forces and the dissipated

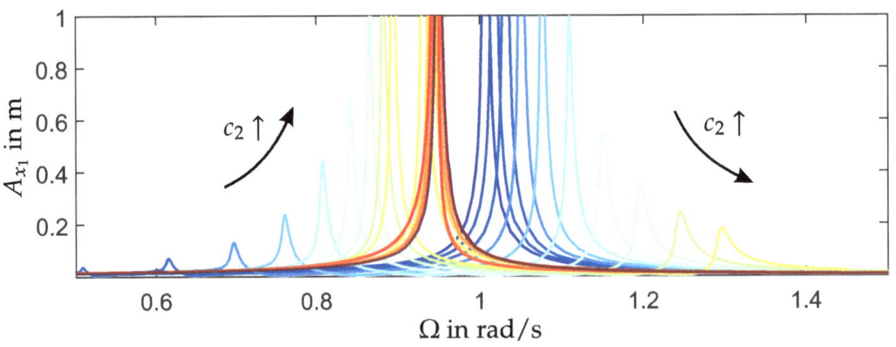

Figure 3.4: Parameter study variation of the stiffness c_2 from $c_2 = 0.01\,\mathrm{N/m}$ (——) to $c_2 = 0.9\,\mathrm{N/m}$ (——).

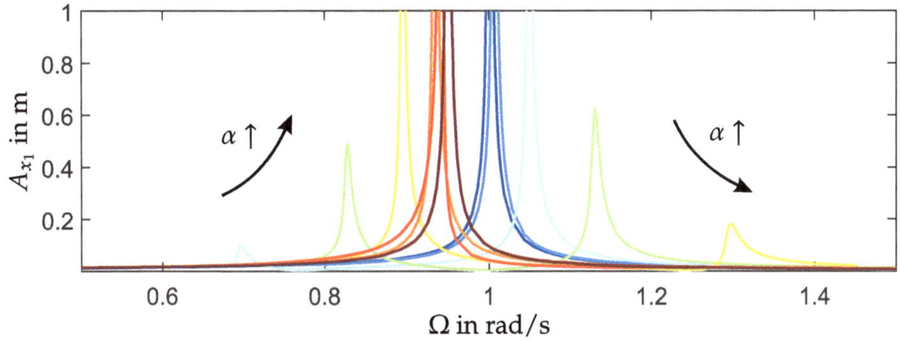

Figure 3.5: Parameter study variation of the angle α from $\alpha = 5°$ (—) to $\alpha = 60°$ (—).

energy. This leads to an initial decrease of the maximum amplitude for small friction coefficients, i.e. $\mu \ll 1$. This decrease is observed as long as the breakaway force does not rise substantially. Large friction coefficients increase the breakaway force and lead to a system that is more likely to stick. Furthermore, large breakaway forces cause a higher limit amplitude A_{L}, c.f. Eq. (3.16). This causes a reduction of the sliding range and two sticking ranges appear at the edges of the frequency response. Thus, the characteristics of the sticking system become dominant. This causes a shift of the system's first resonance peak to the eigenfrequency of the sticking system ω_{st} and higher amplitudes. A transition between the two and one degree of freedom system is observed.

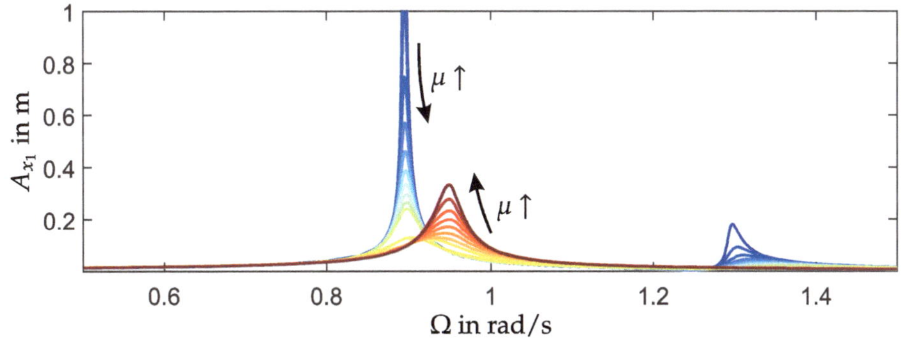

Figure 3.6: Parameter study variation of the friction coefficient μ from $\mu = 0.01$ (—) to $\mu = 1$ (—).

Figure 3.7 presents the variation of the prestress displacement $\Delta\ell$. Similar to the friction coefficient variation, a transition from a two to a one degree of freedom system is observed. The limit amplitude for sticking is increased, and the resonance is shifted towards ω_{st} for increasing prestress displacements. However, a substantial difference lies in the effect on the system, since an increase in the maximum amplitude is observed.

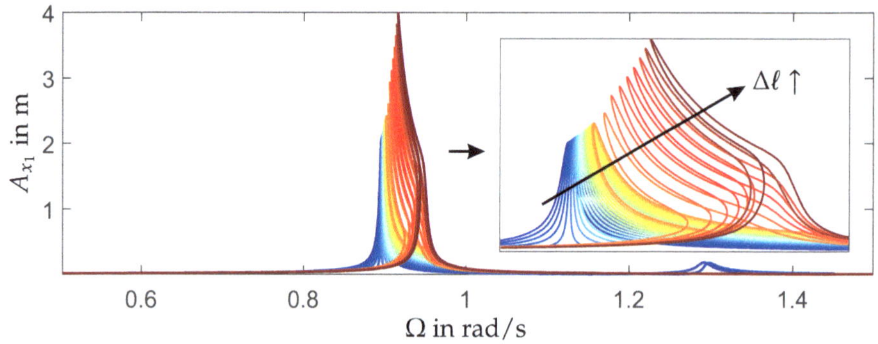

Figure 3.7: Parameter study variation of the prestress displacement $\Delta\ell$ from $\Delta\ell = 0\,\mathrm{m}$ (—) to $\Delta\ell = 1\,\mathrm{m}$ (—).

Based on the work integral, the influence of $\Delta\ell$ on the dissipated energy is different. The prestress displacement increases the dissipated energy associated with the relative amplitude A_{rel}. In contrast, the aforementioned parameters increase the dissipated energy associated with the terms proportional to A_{rel}^2. In consequence, slipping occurs at higher amplitudes without the amplitude reduction observed in the friction coefficient variation. Furthermore, a softening characteristic in the frequency response function is observed. Sticking leads to high amplitudes at the resonance frequency ω_{st}. Once the system transitions into slipping, its resonance frequency changes. Two eigenfrequencies of the slipping system emerge whereas the eigenfrequency of the sticking system vanishes. The lower eigenfrequency of the slipping system influences the damper's dynamics and a softening characteristic develops. The higher the amplitudes the more the softening tends towards the lower eigenfrequency of the slipping system.

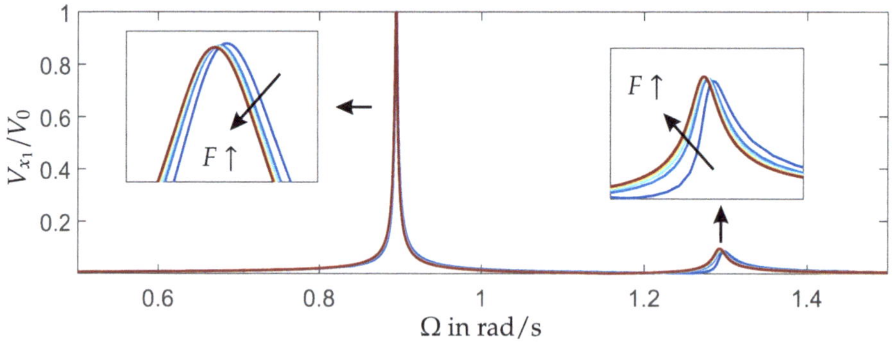

Figure 3.8: Parameter study variation of the force F from $F = 0.01\,\mathrm{N}$ (—) to $F = 0.09\,\mathrm{N}$ (—).

The last parameter variation in Fig. 3.8 considers the change in the harmonic force amplitude F and its effect on the ratio $V_{x_1} = A_{x_1}/F$. In linear systems, this ratio is equivalent to the amplification factor. The curves are normalized by the value V_0, which is the ratio calculated with the standard parameters. As shown in Fig. 3.8 the curves are almost identical. The first resonance peak shows a minimal decrease with an increase in excitation force, whereas the second resonance peak increases. Since these changes are minimal, the vibration amplitudes of the system are amplified by approximately the same factor. This implies a kind of scalability of the amplitude responses for the chosen parameters. Nevertheless, exact scalability is observed only in linear mechanical systems with viscous damping. The scalability is lost with increasing prestress levels. High prestress displacements cause large discrepancies in the nonlinear ranges for varying excitation amplitudes. This is due to the displacement's effect on the limit amplitude. These changes in the nonlinear range cause nonscalable frequency response functions. Nonscalable frequency responses as well as the softening effect noted in the prestress variations are observed in Fig. 3.9. An advantage, however, is that the ratio V_{x_1} does not increase with the excitation force. Instead, the effect noted at the first resonance peak in Fig. 3.8 is amplified. Since the amplification factors of the maximum amplitude do not increase, the system is robust. In contrast, systems with a constant friction force have unbounded amplitudes if the friction force is not tuned to the excitation, c.f. [53].

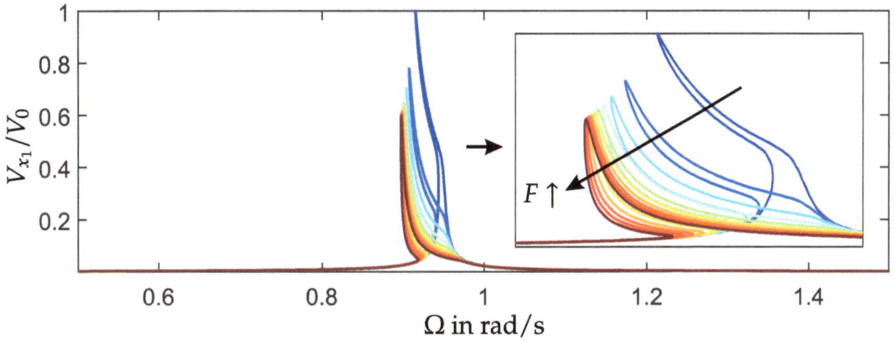

Figure 3.9: Parameter study variation of the force F from $F = 0.01\,\mathrm{N}$ (——) to $F = 0.09\,\mathrm{N}$ (——) with $\Delta\ell = 1\,\mathrm{m}$.

A closer examination of the frequency response function with $\Delta\ell = 1\,\mathrm{m}$ and the remaining standard parameters is presented in Fig. 3.10. A stability analysis is carried out by taking the initial solutions of the harmonic balance method and simulating the system's behavior for 1000 additional cycles. The resulting steady-state solution is compared with the initial periodic solution and the stability is evaluated. Two instability ranges are observed. In the first instability range on the left of Fig. 3.10 multiple solutions are observed, whereas in the second instability range they are not. In the initial instability range, the steady-state solutions drift away from the initial solution and converge either to the upper solution branch or to a nonperiodic steady-state solution. The solutions, however, do not converge to the lower solution branch, where a purely sticking solution is observed. At the second instability range, the steady-state solution

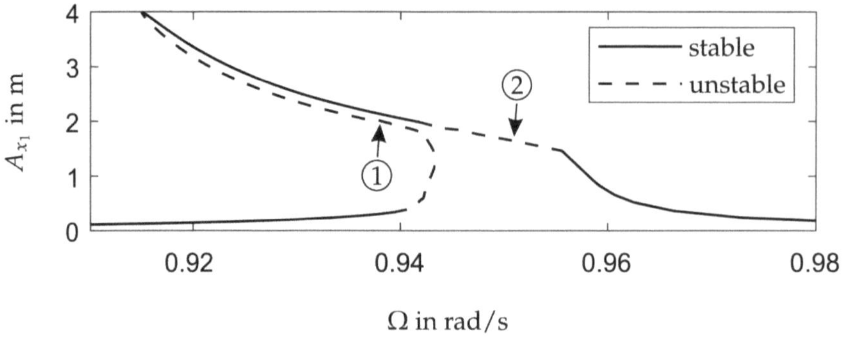

Figure 3.10: Stability analysis of the frequency response function with $\Delta\ell = 1\,\text{m}$.

converges to nonperiodic state solutions. Exemplary solutions from each of these ranges are marked with ① and ② in Fig. 3.10 and are studied in further detail.

To characterize the nonperiodic solutions, time simulations are performed and POINCARÉ maps are constructed, see Fig. 3.11. The states of the system are recorded at multiples of the period $T = 2\pi/\Omega$ and plotted in the $x_1, \dot{x}_1$-phase diagram. The POINCARÉ maps show the transition from the unstable periodic solution to the stable, steady-state, nonperiodic solution. The steady-state solution in Fig. 3.11a corresponds to the first instability range and forms a curve in the phase space. This indicates a quasiperiodic solution. In contrast, the solution from the second instability range in Fig. 3.11b forms an area-like structure. This structure, which is neither a line nor a closed area, is a typical feature of chaos.

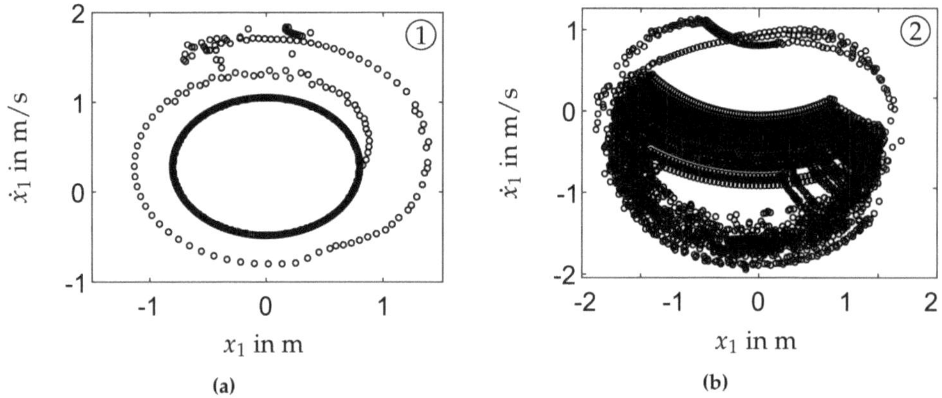

Figure 3.11: (a) Quasiperiodic solution ① at $\Omega = 0.9376\,\text{rad/s}$ with $\Delta\ell = 1\,\text{m}$ and the standard parameters. (b) Chaotic solution ② at $\Omega = 0.9507\,\text{rad/s}$ with $\Delta\ell = 1\,\text{m}$ and the standard parameters.

Unstable solutions are found primarily at high prestress levels. However, they are also found at low prestress levels. A stability analysis for frequency response function of the

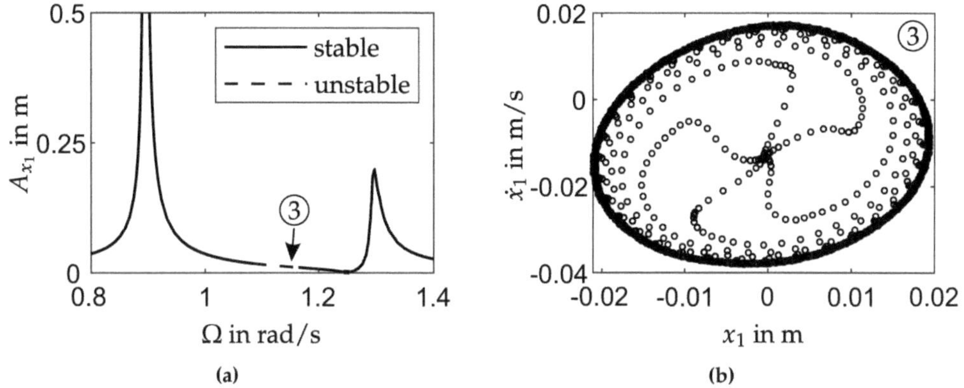

Figure 3.12: (a) Stability analysis of the frequency response function with the standard parameters. (b) Quasiperiodic solution ③ at $\Omega = 1.1446\,\mathrm{rad/s}$ with the standard parameters.

system with the standard parameters as shown in Fig. 3.12a. Unstable periodic solutions at low prestress levels lead to quasiperiodic solutions. An example of such a solution is presented in Fig. 3.12b. Since the maximal amplitudes of nonperiodic solutions are in some cases higher than the calculated periodic motion, these solutions have to be taken into account when designing such dampers and should be avoided.

3.3 Analytical Investigations

The studies above show the influence of the parameters on the frequency response. However, they do not offer a deeper insight into the system's dynamics. Therefore, analytical investigations are carried out with the averaging method described in section 2.2. The analytical solution is calculated and verified with the numerical results. Lastly, the analytical insight is used to optimize the damper parameters.

In this section continuous slip is assumed. The equations of motion during the slip phase are brought into an adequate form for averaging. First, the equations are nondimensionalized. Subsequently, a complete modal decoupling for nonlinear systems is applied. The following transformations are introduced:

$$\begin{gathered}\frac{m_2}{m_1} = \gamma\,,\ \frac{c_1}{m_1} = \omega_{01}^2\,,\ \tau = \omega_{01} t\,,\ \dot{(\ )} = \frac{\mathrm{d}(\ )}{\mathrm{d}t} = \frac{\mathrm{d}(\ )}{\mathrm{d}\tau}\frac{\mathrm{d}\tau}{\mathrm{d}t} = \omega_{01}(\ )'\,,\ \eta = \frac{\Omega}{\omega_{01}}\,,\\ a^2 = \frac{4c_2\tan^2\alpha}{m_1\omega_{01}^2}\,,\ \varepsilon b = \frac{4\mu c_2\tan\alpha}{m_1\omega_{01}^2\cos^2\alpha}\,,\ \varepsilon c = \frac{2c_2\Delta\ell\tan\alpha}{m_1\omega_{01}^2}\,,\ \varepsilon f = \frac{F}{m_1\omega_{01}^2}\,,\ \varepsilon \ll 1\,.\end{gathered} \tag{3.17}$$

The parameter ε tracks the small quantities and is equal to 0.01 for the standard parameters. This implies small friction coefficients, small prestress displacements, and small excitation forces. Linearising the wedge damper force with respect to the friction coefficient and introducing the transformations above yields the coupled nondimensional equations

$$x_1'' + x_1 - a^2(x_2 - x_1) = \varepsilon\big(f \sin \eta\tau + f_{\mathrm{WD,lin}}\big) = \varepsilon f_{\mathrm{WD,lin,1}} \,, \tag{3.18}$$

$$\gamma x_2'' + a^2(x_2 - x_1) = \varepsilon\big(- f_{\mathrm{WD,lin}}\big) = \varepsilon f_{\mathrm{WD,lin,2}} \,, \tag{3.19}$$

$$f_{\mathrm{WD,lin}} = b|x_2 - x_1|\mathrm{sgn}(x_2' - x_1') + c\,\mathrm{sgn}(x_2 - x_1) \,. \tag{3.20}$$

Equations (3.18)–(3.20) only take into account the terms of $O(\varepsilon)$ since only a first order averaging method is applied to the equations. These equations are decoupled and considered in the vicinity of the corresponding eigenfrequency η_{01} or η_{02}. This procedure yields the equations

$$\eta \approx \eta_{01} \quad \rightarrow \quad q_1'' + \eta_{01}^2 q_1 = \varepsilon(r_{11} f_{\mathrm{WD,lin,1}}(q_1, 0) + r_{21} f_{\mathrm{WD,lin,2}}(q_1, 0)) \,, \tag{3.21}$$

$$\eta \approx \eta_{02} \quad \rightarrow \quad q_2'' + \eta_{02}^2 q_2 = \varepsilon(r_{12} f_{\mathrm{WD,lin,1}}(0, q_2) + r_{22} f_{\mathrm{WD,lin,2}}(0, q_2)) \,. \tag{3.22}$$

In Eqs. (3.21) and (3.22) r_{ij} are the corresponding entries in the decoupling modal matrix $\mathbf{R}$ of the unperturbed system, i.e. $\varepsilon = 0$. A Van der Pol transformation of the form $q_i = A_i \sin \varphi_i$, $q_i' = A_i \eta_{0i} \cos \varphi_i$, $\varphi_i = \eta\tau + \theta_i$, and $\varepsilon\delta_i = \eta_{0i} - \eta$ for $i = \{1, 2\}$ is applied. Furthermore, the resulting equations are averaged over one period of the fast oscillation variables φ_1 and φ_2. The averaged equations result in

$$\bar{A}_i' = \varepsilon\left(-\frac{f r_{1i} \sin \bar{\theta}_i}{2\eta_{0i}} - \frac{b \bar{A}_i \, (r_{1i} - r_{2i})^2}{\pi \, \eta_{0i}}\right) , \tag{3.23}$$

$$\bar{\theta}_i' = \varepsilon\left(\delta_i - \frac{\cos \bar{\theta}_i f r_{1i}}{2\bar{A}_i \, \eta_{0i}} + 2\,\frac{c\,(r_{11} - r_{21})\,\mathrm{sgn}\,(r_{1i} - r_{2i})}{\pi \, \bar{A}_i \, \eta_{0i}}\right) . \tag{3.24}$$

Equations (3.23) and (3.24) describe changes in the slow-changing amplitude $\bar{A}_i$ and the slow-changing phase difference $\bar{\theta}_i$ of the system. The approximation of the resonance behavior is approximated by setting $\bar{\theta}_i = -\pi/2$. Although this does not always apply even for linear systems, it is an acceptable assumption since the system is lightly damped. This is verified by considering a phase difference of the form $\theta_i = -\pi/2 + \varepsilon\theta_{i,1}$, where the term $\varepsilon\theta_{i,1}$ represents the slight modification due to light damping. Substituting this expression in the averaged equations and expanding the trigonometric terms with a Taylor series yields terms of the magnitude order $O(\varepsilon^2)$ in Eqs. (3.23) and (3.24). In consequence, the effect of the damping in the phase difference is first noted with an averaging method of second order. Using the proposed phase difference in Eq. (3.23) yields an expression for the maximum amplitude depending on the excitation force. It is derived by solving for the stationary solution $\bar{A}_i' = 0$ and leads to

$$\bar{A}_{i,\max} = \frac{\pi r_{1i} f}{2b(r_{1i} - r_{2i})^2} = \frac{\pi r_{1i} \cos^2 \alpha F}{8\mu c_2 \tan \alpha (r_{1i} - r_{2i})^2} . \tag{3.25}$$

This equation confirms the linear relationship between the maximum amplitude and the amplitude of the excitation force. The pseudo-viscous character of the damper is thus reinforced. Analogously, solving Eq. (3.24) for the resonance case and the stationary solution, i.e. $\bar{\theta}_i' = 0$, leads to a backbone curve. This curve defines the relationship between the maximum amplitude and the resonance frequency. This results in

$$\bar{A}_{i,\max} = \varepsilon \frac{2c|r_{1i} - r_{2i}|}{\pi \eta_{0i}(\eta - \eta_{0i})} . \tag{3.26}$$

An analytical solution is derived for the frequency response function in the resonance regimes from Eqs. (3.23) and (3.24) by eliminating the phase difference θ_i for the stationary solution. This implicit equation is expressed in the original coordinates with the reconstruction rule presented in section 2.2. The comparison between the analytical and the numerical solution is shown in Fig. 3.13. The solution approximates the numerical results with the accuracy of asymptotic methods. The deviations from the numerical solution are of the order of $O(\varepsilon)$, which confirms the validity of the solution. The validity of the backbone curve is also confirmed.

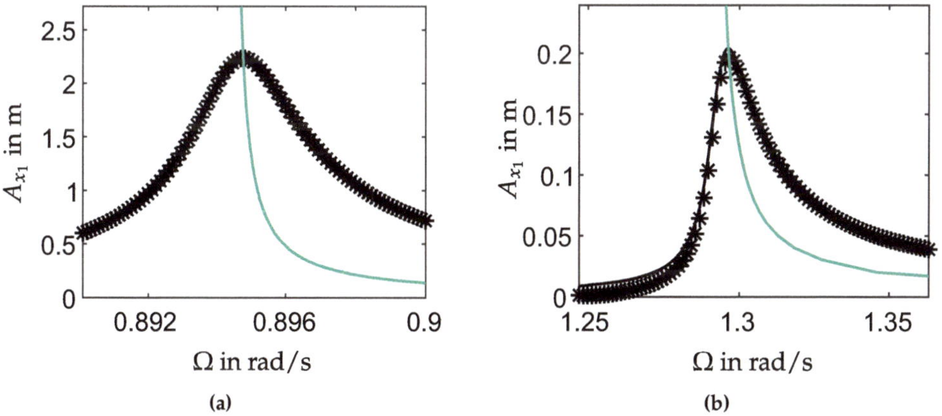

Figure 3.13: Comparison of the analytical (——) and the numerical solution (*), as well as the backbone curve (——). (a) $\Omega \approx \eta_{01}$. (b) $\Omega \approx \eta_{02}$.

The analytic solution assumes continuous slip and thus cannot account for sticking. Consequently, the solution is only applied within the valid parameter range, and the corresponding parameters remain small, i.e. $\mathcal{O}(\varepsilon)$. Nevertheless, the added value of the analytical solution lies in its ability to effectively optimize design functions within its limits. A targeted optimization of the wedge angle α for a broadband damper design is performed. To this end, an optimization function is defined as the sum of the squares of the maximum amplitudes of the main mass at the eigenfrequencies. This leads to

$$g_{\mathrm{opt}}(\alpha) = A^2_{x_1,\mathrm{max}}(\alpha, \eta_{01}(\alpha)) + A^2_{x_1,\mathrm{max}}(\alpha, \eta_{02}(\alpha))\,. \tag{3.27}$$

Equation (3.27) is plotted for the standard parameters in the semi-logarithmic plot in Fig. 3.14a. To ensure the validity of the analytical solution, the wedge angle is limited to $\alpha \in [0°, 45°]$. A minimum of the optimization function is calculated to determine the optimal angle $\alpha_{\mathrm{opt}} = 25.33°$. The resulting frequency response is compared to the response with the standard parameters, see Fig. 3.14b. As with DEN HARTOG's equal peak method, the optimal broadband solution is found when both structural resonances have the same amplitude. Additionally, it achieves an amplitude reduction of 72.5 %.

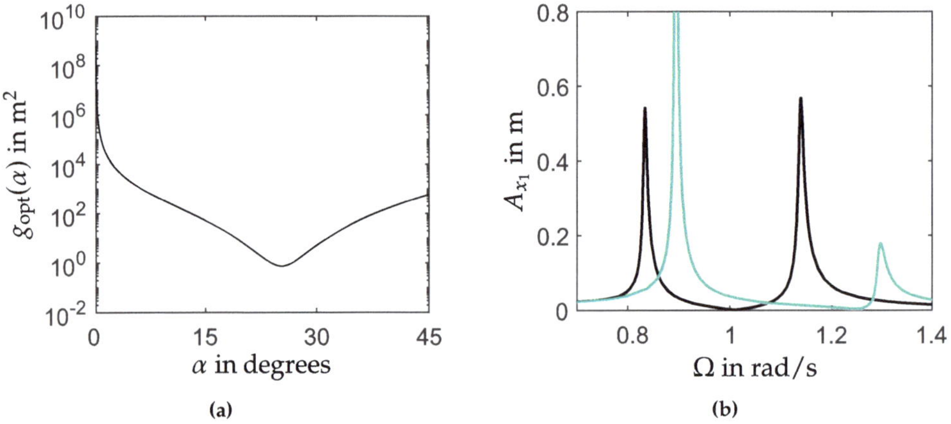

Figure 3.14: (a) Optimization function $g_{\mathrm{opt}}(\alpha)$. (b) Comparison of the optimized frequency response function (——) and the response for the standard parameters (——).

3.4 Damper Assessment

The wedge damper is a realization of a pseudo-viscous damper via dry friction. The dissipated energy of the damper is proportional to the square value of the relative vibration amplitude. This leads to the scalability of the amplitude response for certain parameters. A difference between the wedge damper and the conventional tuned mass damper lies in the proportionality of the dissipated energy. The dissipated energy of the tuned mass damper is proportional to the excitation frequency, whereas that of the wedge damper is not. This is a useful quality at low frequencies. The damper also has the practical advantage that no sealing is needed, which reduces manufacturing costs. Furthermore, the damper is less affected by external changes in the environment in contrast to its viscous counterpart. The damper is in consequence noted as an effective alternative to viscous damping.

The advantages of the wedge damper come at the price of a higher degree of nonlinearity. This eventually leads to quasiperiodic and chaotic solutions, which are detrimental to the system's behavior. Such regimes should be avoided when designing the damper.

The next natural step in the betterment of this damper is the investigation of alternative geometries, such are considered in chapter 4. The extension of this damper into a 3D configuration opens the possibility of independent damping in different directions. An ellipsoidal contact surface with different slopes in the intended damping directions provides a starting point for such investigations.

4 The Tuned Wedge Damper

The tuned wedge damper, see Fig. 4.1, retains the advantages of the wedge damper and addresses its main disadvantage, the coupling of damping and absorption. To this end, the absorption and damping functions are separated by introducing an additional spring and modifying the contact surface design. To study this device, first, the general characteristics of the damper are investigated in section 4.1. Second, the numerical investigations expand the initial insight in section 4.2 by determining the influence of parameters on the system's frequency response function. Stability analyses of selected solutions are also presented. Third, the investigation into the wedge damper is deepened with the analytical consideration of the system in section 4.3. As with the wedge damper, the analytical study leads to relationships between the damper's parameters and key dynamic features, e.g. maximal amplitude. These analytical and numerical insights are combined in section 4.4 where an optimized tuned wedge damper is presented. The chapter is closed in section 4.5 with the damper assessment.

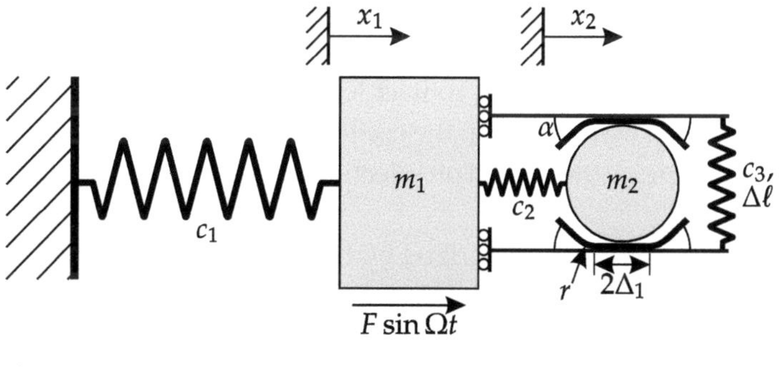

Figure 4.1: The schematic model of the tuned wedge damper.

4.1 General Damper Description

The tuned wedge damper also aims to improve the dynamics of the main system, which is represented by the main mass m_1, the main spring c_1, and the excitation $F \sin \Omega t$. In contrast to the wedge damper, the tuned wedge damper implements two springs and a different contact surface profile. In this way, the damper separates two functions, vibration absorption and vibration damping. The secondary spring c_2 is attached to the secondary mass m_2 and is responsible for vibration absorption. The vibration damping is handled by the tertiary spring c_3 and the contact surfaces. The third spring presses the contact surfaces onto the secondary mass and thus generates a dry friction contact. If additional damping is desired, the third spring is prestressed by the displacement $\Delta\ell$. The contact surface profile is described by three straight segments and two curved connections. The middle segment has a distance of $2\Delta_1$ and is connected to the outer segments with circular segments of radius r. The connections to the outer segments are designed so that the transitions are continuous and smooth. The outer segments are oriented at the wedge angle α. The form of the varying contact geometry is described by the function y, which is specified later on.

The tuned wedge damper is similar to the wedge damper in chapter 3. Both dampers press the secondary mass between varying contact surfaces. Both generate a friction force that is proportional to the relative displacement between the masses. For $c_2 = 0$ and $\Delta_1 = 0$ the wedge damper results as a special case of the tuned wedge damper. Therefore, the tuned wedge damper is an extension of the wedge damper. It is also a more versatile damper since it has more design parameters. Furthermore, its basic working principle is different from the wedge damper. When the oscillations are small, the secondary mass oscillates within the $2\Delta_1$-range. The contact surfaces are not pressed apart, and there is low damping. No damping is achieved in the inner segment for a vanishing prestress displacement ($\Delta\ell = 0$). If the oscillations are larger, the secondary mass oscillates partly outside the $2\Delta_1$-range, and the contact surfaces are pressed apart. This leads to the aforementioned varying contact forces. Moreover, additional damping is introduced in a targeted manner, when the oscillations are critical. The basic principle of the tuned wedge damper is thus based on selective damping with varying contact forces.

The dynamics of the system are described by two differential equations while sliding and by one differential equation while sticking. For the sliding case, these are given by

$$m_1\ddot{x}_1 + c_1 x_1 - F_{\mathrm{TWD}} = F \sin \Omega t \,, \tag{4.1}$$

$$m_2\ddot{x}_2 + F_{\mathrm{TWD}} = 0 \,. \tag{4.2}$$

Again x_1 and x_2 represent the position coordinates of the primary and secondary mass, the dot represents the total derivative with respect to the time t, and F_{TWD} is the horizontal force of the tuned wedge damper. The force of the tuned wedge damper is obtained with the free body diagram presented in Fig. 4.2.

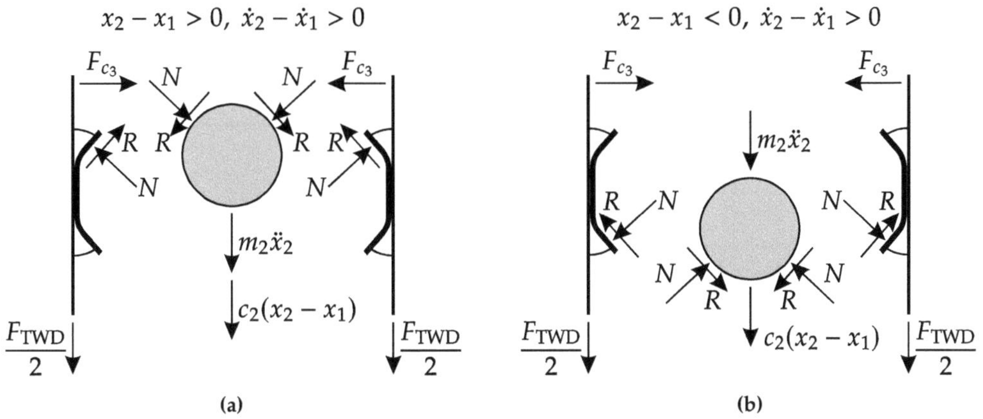

Figure 4.2: Free body diagram of the tuned wedge damper contact forces for: (a) $x_2 - x_1 > 0,\ \dot{x}_2 - \dot{x}_1 > 0$ and (b) $x_2 - x_1 < 0,\ \dot{x}_2 - \dot{x}_1 > 0$.

In contrast to the equations in chapter 3, the momentary angle of the contact force changes depending on the relative displacement $x_{rel} = x_2 - x_1$. To take all segments into account in a general way, the momentary contact angle $\tilde{\alpha}$ is introduced. It is equal to $\pm\alpha$ on the outer segments, zero in the middle segment, and changes from zero to α in the transitions segments. A general expression for the sum of forces is generated by expressing the equations with the tangent of the momentary contact angle. Therefore, the derivative of the contact geometry function with respect to the relative displacement $y_x = \tan(\tilde{\alpha})$ is used. The function y_x includes both cases in Fig. 4.2 and thus sum of forces for the free body diagrams results in

$$R + N y_x - \frac{F_{WD}}{2}\sqrt{1 + y_x^2} = 0\,, \tag{4.3}$$

$$R y_x - N + F_{c_3}\sqrt{1 + y_x^2} = 0\,, \tag{4.4}$$

$$\text{with} \quad R = \mu N \operatorname{sign}(\dot{x}_2 - \dot{x}_1)\,, \quad \text{and} \quad F_{c_3} = c_3(2y + \Delta\ell)\,. \tag{4.5}$$

Since the term $\sqrt{1 + y_x^2} = (\cos(\tilde{\alpha}))^{-1}$ is strictly positive, it limits the considered angles to the range $\alpha \in [-\pi/2, \pi/2]$. However, since the relevant range for practical applications is between 0 and $\pi/2$, this is not a meaningful restriction. While sliding the relationship between the normal force N and the friction force R is given by Coulomb friction and the friction coefficient μ, c.f. Eq. (4.5). Solving the equations above yields the force of the tuned wedge damper in the compact form

$$F_{TWD} = c_2(x_2 - x_1) + 2c_3\,(2y + \Delta\ell)\,\frac{y_x + \mu \operatorname{sgn}(\dot{x}_2 - \dot{x}_1)}{1 - \mu y_x \operatorname{sgn}(\dot{x}_2 - \dot{x}_1)}\,. \tag{4.6}$$

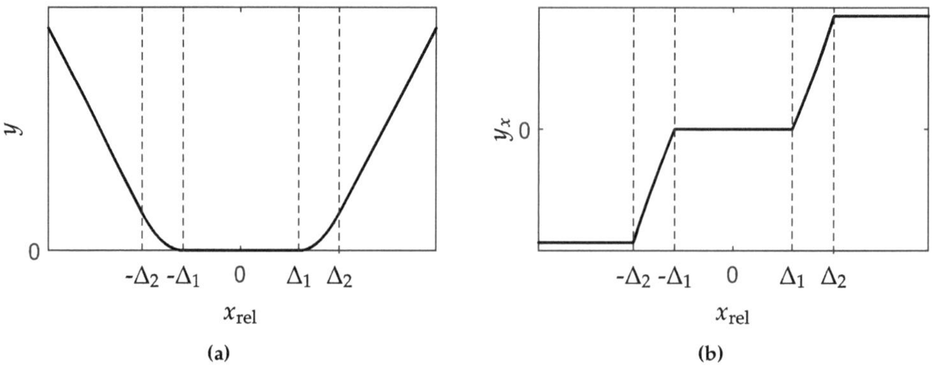

Figure 4.3: (a) Contact function geometry y (b) Derivative of the contact function geometry y_x with respect to the relative displacement x_{rel}.

The transition segments are designed to ensure smooth transitions between the straight segments. To minimize the number of parameters in the design, the transition segments are designed as circular segments. This design has the radius r as its single parameter. Furthermore, this radius is chosen small ($r \ll 1$), so the dynamics of the system are mainly defined by the straight segments. The contact surface geometry y and its derivative y_x result in

$$y = \begin{cases} -\tan\alpha(x_2 - x_1 + \Delta_2) + r(1 - \cos\alpha), & x_2 - x_1 < -\Delta_2 \\ r - \sqrt{r^2 - (x_2 - x_1 + \Delta_1)^2}, & -\Delta_2 < x_2 - x_1 < -\Delta_1 \\ 0, & |x_2 - x_1| < \Delta_1 \\ r - \sqrt{r^2 - (x_2 - x_1 - \Delta_1)^2}, & \Delta_1 < x_2 - x_1 < \Delta_2 \\ \tan\alpha(x_2 - x_1 - \Delta_2) + r(1 - \cos\alpha), & \Delta_2 < x_2 - x_1 \end{cases}, \tag{4.7}$$

$$y_x = \begin{cases} -\tan\alpha, & x_2 - x_1 < -\Delta_2 \\ \dfrac{x_2 - x_1 + \Delta_1}{\sqrt{r^2 - (x_2 - x_1 + \Delta_1)^2}}, & -\Delta_2 < x_2 - x_1 < -\Delta_1 \\ 0, & |x_2 - x_1| < \Delta_1 \\ \dfrac{x_2 - x_1 - \Delta_1}{\sqrt{r^2 - (x_2 - x_1 - \Delta_1)^2}}, & \Delta_1 < x_2 - x_1 < \Delta_2 \\ \tan\alpha, & \Delta_2 < x_2 - x_1 \end{cases}, \tag{4.8}$$

$$\text{with} \quad \Delta_2 = \Delta_1 + r\sin\alpha\,. \tag{4.9}$$

The equation of motion while sticking is identical to the sticking case in chapter 3 and, thus, has the eigenfrequency $\omega_{\mathrm{st}} = \sqrt{c_1/(m_1 + m_2)}$. This equation is given by

$$(m_1 + m_2)\ddot{x}_1 + c_1 x_1 = F\sin\Omega t\ . \tag{4.10}$$

Since the transitions in this system are smooth and the masses are modeled as particles, there is always a defined contact point. Consequently, there is only one sticking condition. The damper force is replaced in the equations of motion with the relation $F_{\mathrm{TWD}} = c_2(x_2 - x_1) - H$, where H describes the stiction force which prevents relative movement. The fulfillment conditions for sticking are described by

$$H = \frac{m_2}{m_1 + m_2}(F \sin \Omega t - c_1 x_1) + c_2(x_2 - x_1)\,, \tag{4.11}$$

$$R = -\frac{2F_{c_3} y_x + H}{2\sqrt{1 + y_x^2}} \quad \text{and} \quad R_{\max} = \mu N = \mu \frac{2F_{c_3} - H y_x}{2\sqrt{1 + y_x^2}}\,, \tag{4.12}$$

$$|R| < R_{\max}\,, \quad \text{and} \quad \dot{x}_1 = \dot{x}_2\,. \tag{4.13}$$

Accordingly, the transition into sliding is thus determined by the failure to comply with the equations above.

The force of the tuned wedge damper is also separated into a conservative part $F_{\mathrm{TWD,c}}$ and a dissipative part $F_{\mathrm{TWD,d}}$. The separation results in

$$F_{\mathrm{TWD,c}} = F_{\mathrm{TWD}}(\mu = 0) = c_2(x_2 - x_1) + 2c_3(2y + \Delta\ell) y_x\,, \tag{4.14}$$

$$F_{\mathrm{TWD,d}} = F_{\mathrm{TWD}} - F_{\mathrm{TWD,c}} = 2c_3(2y + \Delta\ell) \frac{\mu(1 + y_x^2)\,\mathrm{sign}(\dot{x}_2 - \dot{x}_1)}{1 - \mu y_x\,\mathrm{sign}(\dot{x}_2 - \dot{x}_1)}\,. \tag{4.15}$$

The effective secondary stiffness of the damper is defined by the conservative force $F_{\mathrm{TWD,c}}$. In the resonance regime, large amplitudes lead to oscillations that are mainly influenced by the angled segments. The effective secondary stiffness at the resonances is thus given by $c_{2,\mathrm{eff,res}} = c_2 + 4c_3 \tan^2 \alpha$. In contrast, when the oscillations are in the vicinity of the absorption frequency, the relative displacement is within the $2\Delta_1$-range. The angled segments do not influence on the oscillation and the effective secondary stiffness is given by $c_{2,\mathrm{eff,abs}} = c_2$. The structure of the dissipative force is similar to that of the wedge damper. The force is proportional to the relative displacement, i.e. $F_{\mathrm{TWD,d}} \sim A_{\mathrm{rel}}$. Figure 4.4 shows qualitative curves for the tuned wedge damper force and its individual conservative and dissipative parts. The damper forces for $\Delta\ell = 0$ are presented in Fig. 4.4a. This parameter choice leads to a dissipation-free segment in the $2\Delta_1$-range. This is especially useful for absorption. Large prestress displacements lead to the loss of the low dissipation range, see Fig. 4.4b. Furthermore, due to the smooth transitions, the large jumps observed with the wedge damper in chapter 3 are not seen in Fig. 4.4b. This implies that, in comparison to the wedge damper, higher prestress displacements are necessary at $x_{\mathrm{rel}} = 0$ for the tuned wedge damper to stick. Nevertheless, the tuned wedge damper sticks at the angled segments or in the horizontal segment. However,

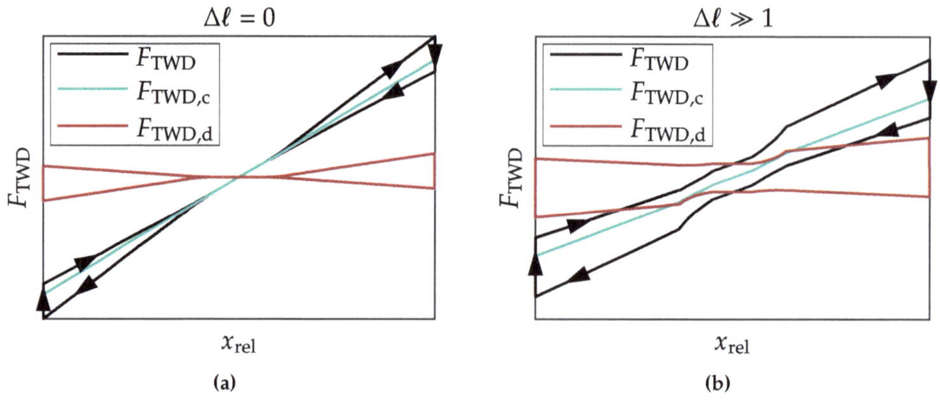

Figure 4.4: Exemplary tuned wedge damper forces for (a) vanishing prestress displacement and (b) large prestress displacement.

sticking in the horizontal segment is the dominant stick-slip transition range. In the middle segment, the full force of the spring force generates the normal force, and the necessary stiction force H defines the contact force R. Since H is an oscillating force, R_{max} is maximized in the horizontal segment. This is verified by setting $y = 0$ and $y_x = 0$ in Eqs. (4.11)–(4.13).

The dissipative work of the damper gives insight into the robustness of the damper and is calculated as with the wedge damper. A harmonic oscillation is assumed for the relative coordinate, i.e. $x_{rel} = A_{rel} \sin \varphi$ and $\dot{x}_{rel} = A_{rel}\Omega \cos \varphi$ with $\varphi = \Omega t$. The work integral is evaluated for two cases. First, oscillations in the $2\Delta_1$-range are considered, which leads to $W_{F_{TWD,d},\Delta_1}$. Subsequently, oscillations in which the secondary mass partly enters the outer segments are handled and lead to the dissipated energy $W_{F_{TWD,d}}$. For small oscillation, the work integral yields

$$W_{F_{TWD,d},\Delta_1} \approx -\int F_{TWD,d}(x_{rel}, \dot{x}_{rel})\,\mathrm{d}x_{rel} = -8c_3\Delta\ell\mu A\,. \tag{4.16}$$

From Eq. (4.16) the dissipation for small oscillations is dependent on the prestress displacement. If this value is null, there is no dissipation in this segment. Furthermore, the energy dissipated in this segment is only proportional to the value of the relative amplitude, i.e. $W_{F_{TWD,d},\Delta_1} \sim A_{rel}$. This structure leads to amplitudes larger than Δ_1 if the presstress is not adjusted to the excitation amplitude F, c.f. [187]. Since in general $\Delta\ell$ is not adjusted for F, the relative amplitudes will in most cases be larger than Δ_1. This is especially true in the resonance regimes. To simplify the calculations of the dissipated energy for the second case, the radius r of the transitions segments is set to null. This is not a strong restriction, since the transition segments are small and only functions to ensure smooth transitions. Only the straight segments are considered and the integral is evaluated for only for these segments. The work integral yields

$$\begin{aligned}
W_{F_{\mathrm{TWD,d}}} &\approx -\int F_{\mathrm{TWD,d}}(x_{\mathrm{rel}}, \dot{x}_{\mathrm{rel}})\,\mathrm{d}x_{\mathrm{rel}}\,, \\
W_{F_{\mathrm{TWD,d}}} &\approx -8\mu c_3 \frac{1+\tan^2\alpha}{1-\mu^2\tan^2\alpha}(\tan\alpha\, A_{\mathrm{rel}}^2 + \Delta\ell A_{\mathrm{rel}}) \\
&\quad + 8\mu c_3 \frac{1+\tan^2\alpha}{1-\mu^2\tan^2\alpha}\left(2A_{\mathrm{rel}} + \Delta\ell\sin\alpha\cos\alpha(1+\mu^2) - \Delta_1\right)\tan\alpha\Delta_1\,.
\end{aligned} \tag{4.17}$$

The first part of Eq. (4.17) is identical to the dissipated work of the wedge damper. They are in part proportional to the square value of the relative displacement amplitude, i.e. $W_{F_{\mathrm{TWD,d}}} \sim A_{\mathrm{rel}}^2$. This ensures the damper's robustness with respect to the excitation amplitude. The second term is in part proportional to the distance Δ_1 and in part to its square value Δ_1^2. If Δ_1 is set to zero the dissipated energy of the damper is identical to that of the wedge damper. The Δ_1-terms reduce the dissipated energy. However, since there are dissipative terms are proportional to the square value of the relative displacement and independent of Δ_1^2, the reduction does not affect the damper's robustness.

The considered dampers have similar characteristics in the resonance regimes. With dissipated energy proportional to the square value of the relative amplitude both dampers are robust against large oscillations. Both dampers also have a dissipated energy independent of the excitation frequency. This is advantageous at low frequencies and does not introduce exceedingly high damping at high frequencies. The dampers differ for small oscillations since the tuned wedge damper exhibits no dissipated energy in this regime for a vanishing $\Delta\ell$ value.

The transition between sticking and sliding is also approximated. To this end, sticking is assumed as well as a harmonic oscillation of the main mass, i.e. $x_1 = A\sin\varphi$. Since the stick-slip transition occurs mainly in the horizontal segment, y and y_x are set to zero. The approximation for the adherence to the sticking conditions results in

$$\begin{aligned}
|R(y=0, y_x=0)| &= \frac{|H|}{2} = \frac{|m_2\ddot{x}_1 + c_2 x_{\mathrm{rel}}|}{2}\,, \\
R_{\max}(y=0, y_x=0) &= \mu N(y=0, y_x=0) = \mu F_{c_3} = \mu c_3 \Delta\ell\,. \\
|R(y=0, y_x=0)| &= \frac{|m_2\ddot{x}_1 + c_2 x_{\mathrm{rel}}|}{2} \leq \frac{m_2|\ddot{x}_1| + c_2|x_{\mathrm{rel}}|}{2} = \frac{m_2\Omega^2 A|\sin\varphi| + c_2|x_{\mathrm{rel}}|}{2} \\
&\leq \frac{m_2\Omega^2 A + c_2|x_{\mathrm{rel}}|}{2} \leq \mu N(y=0, y_x=0) = \mu c_3\Delta\ell\,.
\end{aligned}$$

Solving the last inequality for the amplitude A yields the estimate

$$A \leq \frac{2\mu c_3 \Delta\ell - c_2|x_{\mathrm{rel}}|}{m_2\Omega^2} = A_{\mathrm{L}}\,. \tag{4.18}$$

Since the triangle inequality is used, Eq. (4.18) is a conservative estimate. Consequently, all amplitudes below the limit curve A_L are of the linear system, whereas amplitudes well above the limit curve are of the nonlinear system. Although amplitudes slightly above the limit curve may still be linear.

4.2 Numerical Investigations

The numerical investigations expand the initial insight of the tuned wedge damper. This section is divided into two parts. This first one considers parameter variations and their effect on the frequency response function. The second part of this section focuses on selected frequency responses. Stability analyses are carried out, and POINCARÉ maps are constructed. The standard parameters are chosen with a vanishing prestress displacement, to focus on the decoupling of absorption and damping characteristics of the damper. Unless mentioned otherwise, the standard parameters of the tuned wedge damper are

$$m_1 = 1\,\mathrm{kg},\ m_2 = 0.1\,\mathrm{kg},\ c_1 = 1\,\mathrm{N/m},\ c_2 = 0.1\,\mathrm{N/m},\ c_3 = 0.01\,\mathrm{N/m},\ \alpha = 20^\circ,$$
$$\Delta\ell = 0\,\mathrm{m},\ \Delta_1 = 0.1\,\mathrm{m},\ r = 0.01\,\mathrm{m},\ \mu = 0.1,\ F = 0.01\,\mathrm{N}\,.$$

The variations of the tertiary stiffness c_3 and the outer segment angle α are considered first. The effects of these parameters are similar to each other and are presented in Figs. 4.5 and 4.6. These parameters influence both the amplitude and the position of the

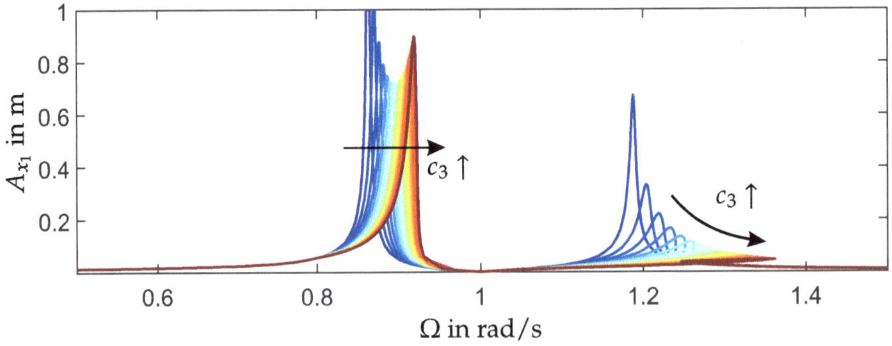

Figure 4.5: Parameter study variation of the stiffness c_3 from $c_3 = 0.01\,\mathrm{N/m}$ (——) to $c_3 = 0.2\,\mathrm{N/m}$ (——).

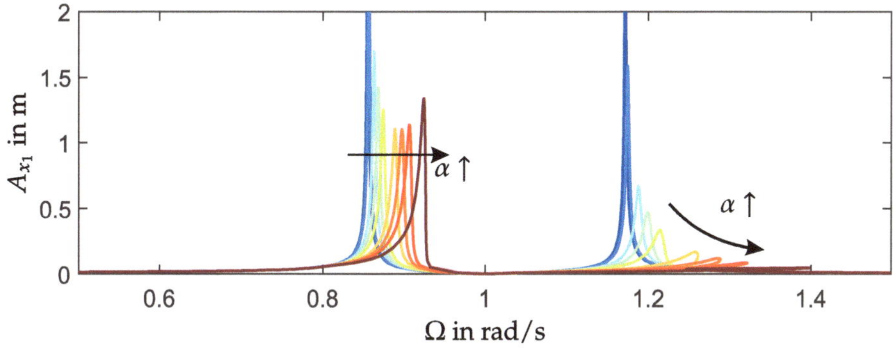

Figure 4.6: Parameter study variation of the stiffness α from $\alpha = 0°$ (——) to $\alpha = 60°$ (——).

peaks. The absorption frequency, however, remains unaffected. The influence of these parameters is also smaller in comparison to chapter 3. This is due to the structure of the system. Since these parameters influence the effective secondary stiffness for large oscillations and not at the absorption frequency, their influence is limited to the resonance regimes. Furthermore, these parameters modify the effective secondary stiffness in the resonance only partly, since this stiffness is also determined by the secondary spring c_2. The increase in the effective secondary stiffness caused by these parameters leads to higher eigenfrequencies. Additionally, optimal values for the stiffness c_3 and the angle α are noted for the peak of the first resonance frequency. In contrast, the amplitudes of the second resonance peak are reduced with increasing values of c_3 and α.

The variation of the parameter Δ_1, which determines the length of the horizontal segment, is presented in Fig. 4.7. An increase of the Δ_1-value leads to higher amplitudes at both peaks. The increase of the resonance amplitudes is caused by the decrease

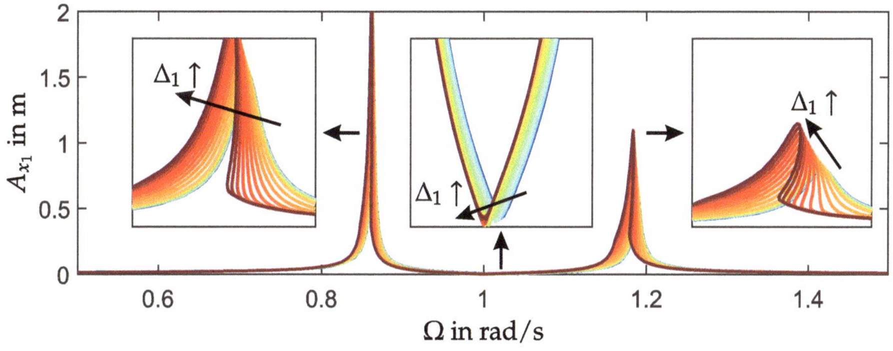

Figure 4.7: Parameter study variation of the Δ_1-range from $\Delta_1 = 0.001$ m (——) to $\Delta_1 = 1$ m (——).

in dissipated energy due to the Δ_1-terms, c.f. Eq. (4.17). Furthermore, if the Δ_1-value is increased enough frequency ranges with multiple solutions are observed. At low vibration amplitudes only the stiffness $c_{2,\text{eff,abs}} = c_2$ is active, whereas for exceedingly high oscillations the stiffness is approximately given by $c_{2,\text{eff,res}} = c_2 + 4c_3 \tan^2 \alpha$. This stiffening in combination with high amplitudes leads to a range with coexisting solutions. Additionally, the location of the peaks shifts towards lower frequencies. An increase of the horizontal segment's length leads to a larger range where only the stiffness $c_{2,\text{eff,abs}}$ is dominant. Because of this $c_{2,\text{eff,res}}$ is less active and the system's stiffness is reduced in comparison to lower Δ_1-values. In addition, a positive effect is noted at the absorption frequency. For longer horizontal segments lower absorption amplitudes are noted.

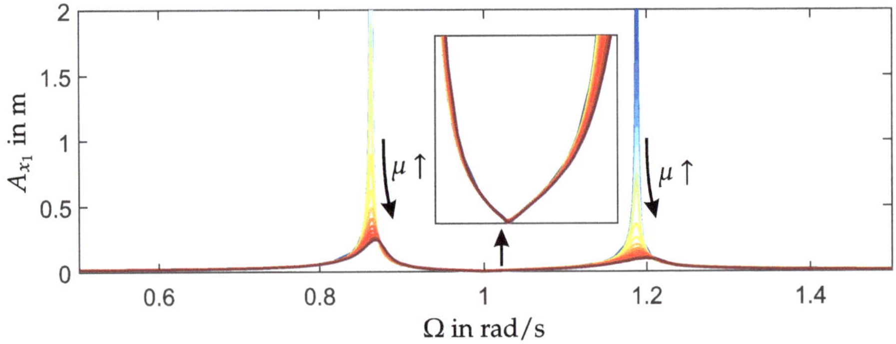

Figure 4.8: Parameter study variation of the friction coefficient μ from $\mu = 0.01$ (——) to $\mu = 1$ (——).

Figure 4.8 depicts the variations of the friction coefficient μ. The decoupling of damping and absorption is observed. An increase of the friction coefficient leads to lower resonance peaks while the absorption remains unaffected. This is caused by the selective damping introduced by the damper. Small vibration amplitudes oscillate within the horizontal segment, thus little to no damping is introduced. The damper acts as a vibration absorber and is able to counter the vibrations of the main system. In the resonance regime, where the vibrations are larger, the angled segments come into play and introduce damping. This behavior is observed at low prestress levels, where minimal damping is introduced in the horizontal segment. If the prestress is increased, the decoupling effect is impaired, see Fig. 4.9. In this case high friction coefficients introduce excessive damping and the decoupling and absorption are lost. A transition into a one degree of freedom system is observed. In contrast to the wedge damper, the shift towards the sticking frequency is less pronounced. The resonance of the sticking system remains in the vicinity of the first eigenfrequency for the chosen parameters.

The variations of the prestress displacement $\Delta\ell$ are presented in Fig. 4.10. An increase in this parameter leads to an increase in the sticking ability of the system. An increase

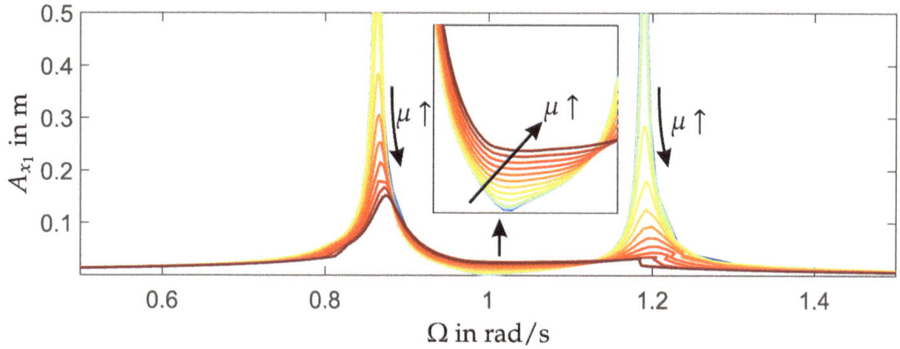

Figure 4.9: Parameter study variation of the friction coefficient μ from $\mu = 0.01$ (—) to $\mu = 1$ (—) with $\Delta\ell = 0.01$ m.

also leads to lower amplitudes since the dissipated energy increases. However, if the prestress is set high, the system sticks excessively and less energy is dissipated. Therefore, for high prestress displacements the resonance amplitudes rise again. Both eigenfrequencies vanish and the system tends to the resonance frequency of the sticking system. Like the wedge damper, a softening effect is observed. This again occurs due to the change in eigenfrequencies caused by the transition from sticking to sliding. Once the system slips, the sticking eigenfrequency vanishes, and the eigenfrequencies of the two degrees of freedom system emerge. The system then tends towards the slipping system's first eigenfrequency and a softening effect is observed. In contrast to the wedge damper, the amplitudes do not rise at first. This implies that the transition point from sticking to sliding is better tuned to its dissipative capabilities than the simple wedge damper. Lastly, the increase of the prestress displacement introduces dissipation in the horizontal segment. Thus no absorption is observed for high $\Delta\ell$-values.

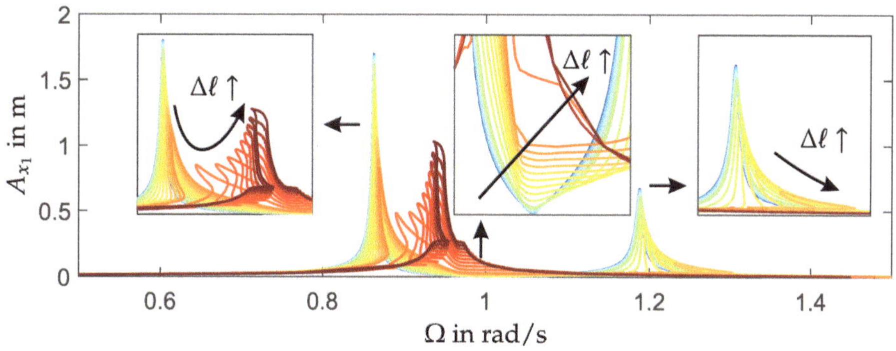

Figure 4.10: Parameter study variation of the prestress displacement $\Delta\ell$ from $\Delta\ell = 0.001$ m (—) to $\Delta\ell = 10$ m (—).

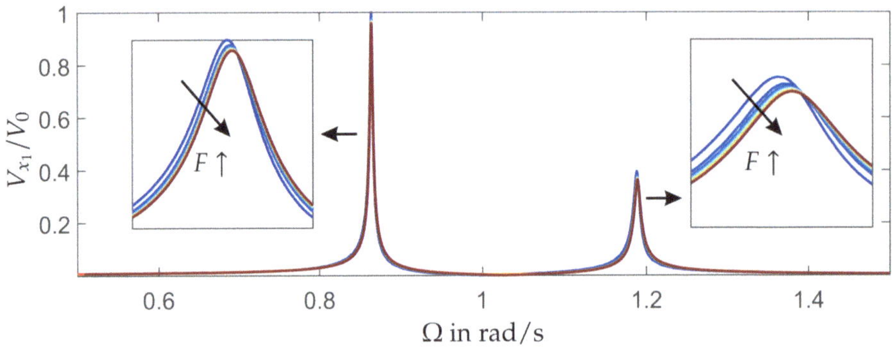

Figure 4.11: Parameter study variation of the force F from $F = 0.01\,\mathrm{N}$ (——) to $F = 1\,\mathrm{N}$ (——).

The last parameter variations focus on the effect of the excitation on the ratio $V_{x_1} = A_{x_1}/F$. The ratio V_{x_1} is normalized by the value V_0 which is the value of V_{x_1} for the standard parameters. In Fig. 4.11 the curves lie nearly over each other and the maximum difference between the curve peaks is of 4.23 % with respect to the standard parameters. Thus, the system shows an almost scalable response for the standard parameters. This confirms the damper's robustness with respect to the excitation. The ratio V_{x_1} does not increase indefinitely as is the case with friction dampers that deliver a constant friction force. In Fig. 4.12, a similar behavior as with the wedge damper is observed for a high prestress value. The scalability of the frequency response function is lost. In contrast to the wedge damper, the amplification factor rises. However, not unbounded as is the case with conventional friction dampers. Instead a saturation is observed as the system tends towards a two degree of freedom system. The saturation is due to the decreasing influence of the horizontal segment. For high excitation forces the dissipation in the $2\Delta_1$-range is nearly irrelevant, whereas for low excitation forces it is the dominant dissipation mechanism.

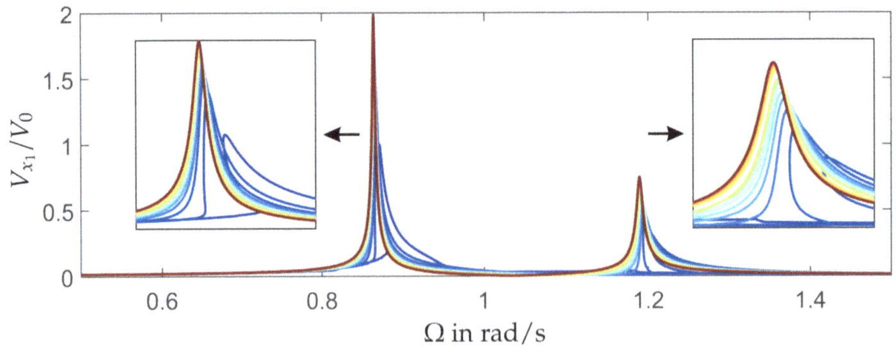

Figure 4.12: Parameter study variation of the force F from $F = 0.01\,\mathrm{N}$ (——) to $F = 1\,\mathrm{N}$ (——) with $\Delta\ell = 1\,\mathrm{m}$.

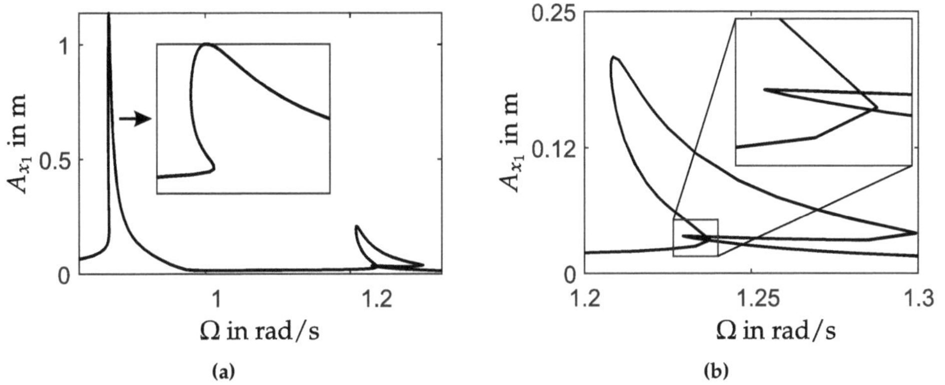

Figure 4.13: Frequency response function for $\Delta\ell = 0.6\,\mathrm{m}$ for: (a) $\Omega \in [0.5, 1.5]$ and (b) $\Omega \in [1.2, 1.3]$.

The rest of the numerical investigations focuses, in an exemplary manner, on the system's solution with $\Delta\ell = 0.6\,\mathrm{m}$. In addition to the softening behavior at the first eigenfrequency, see Fig. 4.13a, at the second eigenfrequency the solution folds three times, appears to cross itself, and folds again, see Fig. 4.13b. To further examine this behavior, additional information is considered. To this end, the phase difference θ_{x_1} of the oscillation with respect to the excitation is taken into account. Since the system is nonlinear, the phase difference is not necessarily constant over an oscillation period. Therefore, a reference point for the phase different is chosen. For the sake of simplicity, the phase difference at $t = 0$ is considered. However, any other time point can be chosen, as long as the reference point remains constant for all the solutions. Assuming an oscillation of the form $x(t) = A_{x_1} \sin(\Omega t + \theta_{x_1}(t))$ and $\dot{x}(t) = A_{\dot{x}_1} \cos(\Omega t + \theta_{x_1}(t))$, the phase difference is calculated with

$$\theta_{x_1}(t = 0) = \arctan\left(\frac{x(t = 0)/A_{x_1}}{\dot{x}(t = 0)/A_{\dot{x}_1}}\right). \tag{4.19}$$

A similar approach was presented by Sokolov and Babitsky concerning a system with a cubic stiffness [159]. The nature of the apparent intersection becomes clear in Fig. 4.14a. In the 3D perspective, no intersection is observed. The crossing points in Fig. 4.13b are equal in amplitude and frequency, however, they differ in phase difference. Instead, these intersections are caused by four bends and the two-dimensional projection of the 3D curve. Furthermore, two projections are derived from Fig. 4.14a and are presented in Figs. 4.14b and 4.14c. The projection in Fig. 4.14b presents the phase difference θ_{x_1} of the main mass at $t = 0$ with respect to the excitation frequency. The outer ranges of the Ω-θ_{x_1}-curve are similar to a linear system since it starts at 0 and ends at π. However, it differs in its transition due to the multiple bends caused by the nonlinearities in the resonance regimes. The ranges with coexisting solutions for a single frequency are also observed. Analogously, in Fig. 4.14b the crossing is a result of the 2D projection of the 3D curve. Figure 4.14c relates the amplitude and phase difference of the coordinate x_1. It is noted that the maximum amplitude occurs approximately at $-\pi/2$. This corroborates the assumptions for the subsequent analytical investigations at the resonance in section 4.3.

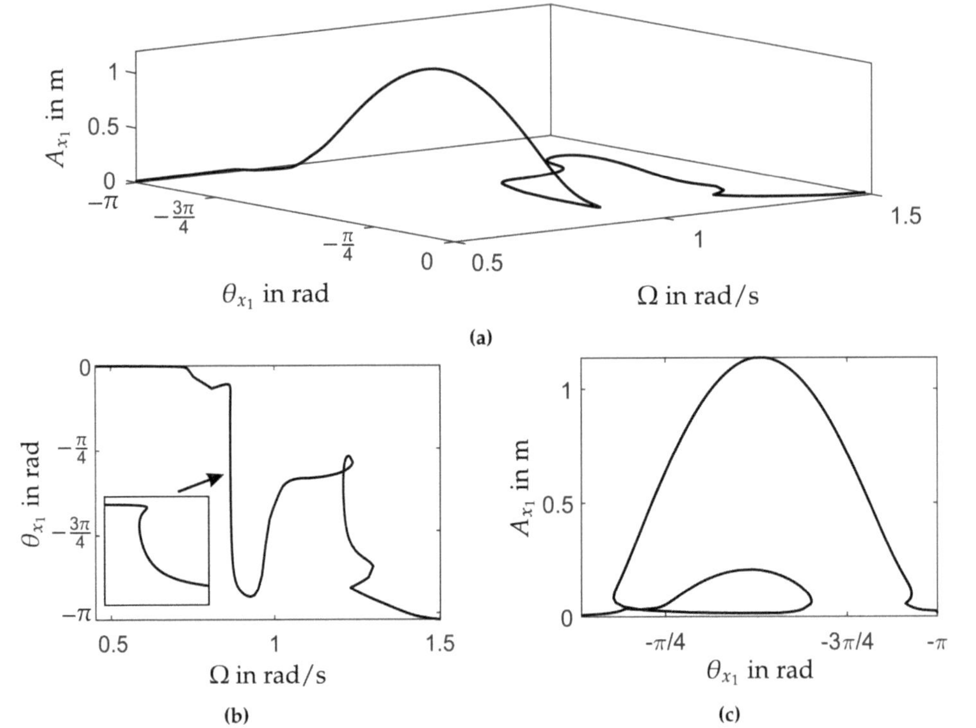

Figure 4.14: Response function for $\Delta\ell = 0.6$ m for: (a) Ω v.s. θ_{x_1} v.s. A_{x_1}, (b) Ω v.s. θ_{x_1}, and (c) θ_{x_1} v.s. A_{x_1}.

The stability investigations are carried out by simulating 1000 additional cycles starting from the calculated solution. If the solution diverges from the calculated solution it is considered unstable. Otherwise it is considered stable. The results of these simulations are shown in Fig. 4.15. Four instability ranges are noted. The first occurs in the softening region with multiple solutions of the first eigenfrequency. The second instability range is found in between both eigenfrequencies. The third and fourth ranges occur due to the bends in the vicinity of the second eigenfrequency. Exemplary solutions from each of these ranges are marked with ①, ②, ③, and ④ and are investigated further in detail.

For the chosen unstable solutions Poincaré maps are constructed. The maps are constructed by recording the states at subsequent time intervals of the excitation period $T = 2\pi/\Omega$. The results are presented in Fig. 4.16. The first unstable solution in Fig. 4.16a

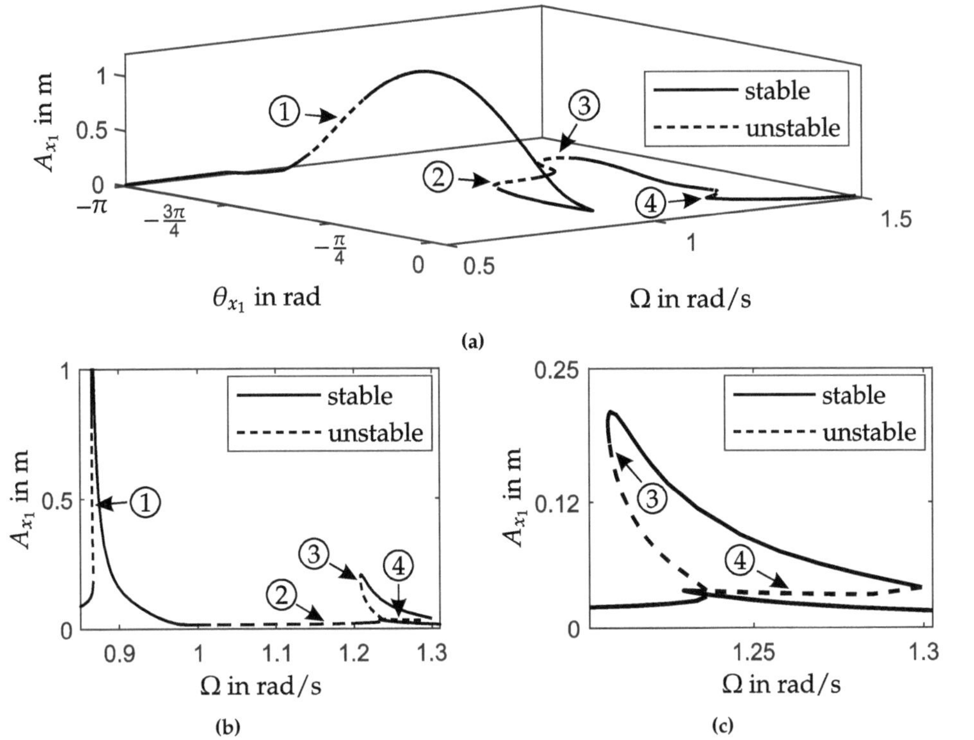

Figure 4.15: Stability analysis of the frequency response function with $\Delta\ell = 0.6\,\text{m}$: (a) 3D-perspective, (b) 2D-perspective, and (c) 2D-perspective zoom of the second eigenfrequency.

converges to the periodic solution of the upper branch. As with the wedge damper, the solutions in this instability range does not converge to the sticking solution of the lower branch. Its Poincaré map is reduced to a curved segment which shows the transient behavior from the unstable to the sable solution. The stable periodic solution is reduced to a single point in the Poincaré map. The solution of the second instability range is shown in Fig. 4.16b and converges to a quasiperiodic solution. This is noted by the closed curve in the Poincaré map. The chosen quasiperiodic solution is more involved than the quasiperiodic solutions of the wedge damper and crosses itself on two occasions. However, not all solutions in this instability range converge to quasiperiodic solutions with intersections. The last solutions in the third and fourth instability ranges are presented in Figs. 4.16c and 4.16d present the same behavior as the first solution. The solutions start from an unstable periodic solution, converge to a stable periodic solution, and do not stick. Their Poincaré maps are, therefore, reduced to a curve. Throughout the instability ranges no chaotic solutions were observed.

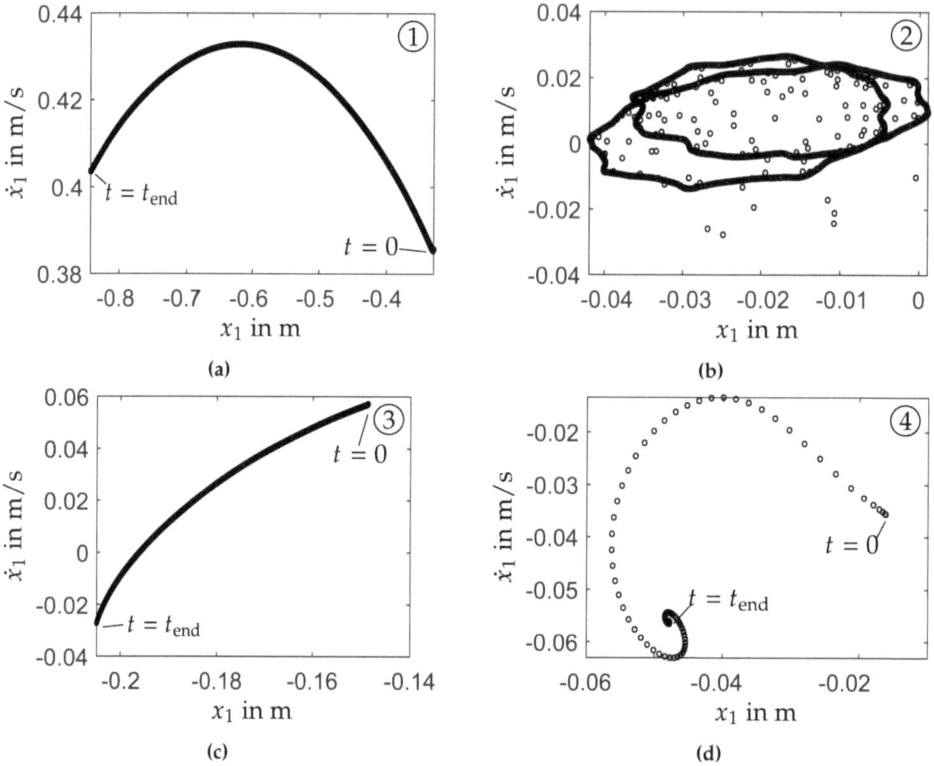

Figure 4.16: Poincaré maps of the unstable solutions at different frequencies: (a) ① $\Omega = 0.865\,\text{rad/s}$ (b) ② $\Omega = 1.196\,\text{rad/s}$, (c) ③ $\Omega = 1.209\,\text{rad/s}$, and (d) ④ $\Omega = 1.264\,\text{rad/s}$.

4.3 Analytical Investigations

The understanding of the tuned wedge damper dynamics is further expanded via the analytical methods presented in section 2.2. To this end, continuous slip is assumed, and only the equations of motion while sliding are considered. First, the equations are nondimensionalized. Second, the equations are brought in standard form for averaging with a modal transformation and a modal decoupling. Once the equations are in proper form, the averaging method is applied and the validity of the solution is verified by a comparison with the numerical solution. Furthermore, relationships between the system's parameters and key dynamic features are derived. These include the maximum amplitude depending on the system parameters and the excitation amplitude. Additionally, an expression for the system's backbone curve is also derived. These relationships are later used to optimize the system's frequency response.

To nondimensionalize the system's equations of motion the following transformations are introduced

$$\begin{aligned}
&\frac{m_2}{m_1} = \gamma\,,\ \frac{c_1}{m_1} = \omega_{01}^2\,,\ \tau = \omega_{01} t\,,\ \dot{(\ )} = \frac{\mathrm{d}(\ )}{dt} = \frac{\mathrm{d}(\ )}{d\tau}\frac{\mathrm{d}\tau}{dt} = \omega_{01}(\ )'\,,\ \eta = \frac{\Omega}{\omega_{01}}\,,\\
&p^2 = \frac{c_2}{c_1}\,,\ \varepsilon a = \frac{4c_3}{m_1\omega_{01}^2}\,,\ \varepsilon b = \frac{2c_3\Delta\ell}{m_1\omega_{01}^2}\,,\ \varepsilon c = \frac{4c_3\mu}{m_1\omega_{01}^2}\,,\ \varepsilon d = \frac{2c_3\Delta\ell\mu}{m_1\omega_{01}^2}\,,\\
&\varepsilon f = \frac{F}{m_1\omega_{01}^2}\,,\ r = O(\varepsilon)\,,\ \varepsilon \ll 1\,.
\end{aligned} \tag{4.20}$$

The parameter ε tracks the small quantities of the system. This implies a soft tertiary spring and small excitation forces. Additionally, the transition radius is also of the magnitude order ε. In contrast, the friction coefficient is also small, however, not of the magnitude order ε. This allows a linearisation of the damper force with respect to the friction coefficient. The equations with the linearised force are given with the transformations above and result in

$$x_1'' + x_1 - p^2(x_2 - x_1) = \varepsilon(f\sin\eta\tau + f_{\mathrm{TWD,lin}}) = \varepsilon f_{\mathrm{TWD,lin,1}}\,, \tag{4.21}$$

$$\gamma x_2'' + p^2(x_2 - x_1) = -\varepsilon f_{\mathrm{TWD,lin}} = \varepsilon f_{\mathrm{TWD,lin,2}}\,, \tag{4.22}$$

$$f_{\mathrm{TWD,lin}} = a y y_x + b y_x + c y(y_x^2 + 1)\mathrm{sgn}(x_2' - x_1') + d(y_x^2 + 1)\mathrm{sgn}(x_2' - x_1')\,. \tag{4.23}$$

In order to accurately approximate the system at its resonance frequency, the stiffness caused by the angled segments is added to both sides of the equations. The introduction of these terms provides a more accurate description of the underlying linear system at the resonance. The modified equations are given by

$$\begin{aligned}
x_1'' + x_1 - p^2(x_2 - x_1) - 4a\tan^2\alpha(x_2 - x_1)&\\
= \varepsilon(f\sin\eta\tau + \tilde{f}_{\mathrm{TWD,lin}}) &= \varepsilon\tilde{f}_{\mathrm{TWD,lin,1}}\,,
\end{aligned} \tag{4.24}$$

$$\gamma x_2'' + p^2(x_2 - x_1) + 4a\tan^2\alpha(x_2 - x_1) = -\varepsilon\tilde{f}_{\mathrm{TWD,lin}} = \varepsilon\tilde{f}_{\mathrm{TWD,lin,2}}\,, \tag{4.25}$$

$$\begin{aligned}
\tilde{f}_{\mathrm{TWD,lin}} =& a y y_x + b y_x + c y(y_x^2 + 1)\mathrm{sgn}(x_2' - x_1') + d(y_x^2 + 1)\mathrm{sgn}(x_2' - x_1')\\
&- 4a\tan^2\alpha(x_2 - x_1)\,.
\end{aligned} \tag{4.26}$$

A modal transformation is applied to Eqs. (4.24) and (4.25). In addition, only the corresponding modal coordinate is considered in the vicinity of the eigenfrequencies η_{01} and η_{02}. Again the values r_{ij} correspond to the entries of the decoupling modal matrix $\mathbf{R}$ of the unperturbed system, i.e. $\varepsilon = 0$. The decoupling results in equations of the form

$$\eta \approx \eta_{01} \quad \rightarrow \quad q_1'' + \eta_{01}^2 q_1 = \varepsilon(r_{11}\tilde{f}_{\mathrm{TWD,lin,1}}(q_1, 0) + r_{21}\tilde{f}_{\mathrm{TWD,lin,2}}(q_1, 0))\,, \tag{4.27}$$

$$\eta \approx \eta_{02} \quad \rightarrow \quad q_2'' + \eta_{02}^2 q_2 = \varepsilon(r_{12}\tilde{f}_{\mathrm{TWD,lin,1}}(0, q_2) + r_{22}\tilde{f}_{\mathrm{TWD,lin,2}}(0, q_2))\,. \tag{4.28}$$

Applying a Van der Pol transformation of the form $q_i = A_i \sin\varphi_i$, $q_i' = A_i\eta_{0i}\cos\varphi_i$, $\varphi_i = \eta\tau + \theta_i$, and $\varepsilon\delta_i = \eta_{0i} - \eta$ for $i = \{1, 2\}$, the equations are brought into standard form and subsequently averaged. The averaged equations for the slowly varying amplitudes $\bar{A}_i$ and the slow varying phase difference $\bar{\theta}_i$ are given by

$$\begin{aligned}\bar{A}_i' = \varepsilon\Bigg(& -\frac{r_{1i} f \sin\bar{\theta}_i}{2\eta_{0i}} \pm \frac{4c\tan\alpha\left(\bar{A}_i(r_{1i} - r_{2i}) \pm \Delta_1\right)^2 \operatorname{sgn}(r_{1i} - r_{2i})}{\pi\cos^2\alpha\bar{A}_i\eta_{0i}} \\ & \mp \frac{4d\left(\bar{A}_i(r_{1i} - r_{2i}) \pm \Delta_1 \sin^2\alpha\right)\operatorname{sgn}(r_{1i} - r_{2i})}{\pi\cos^2\alpha\bar{A}_i\eta_{0i}}\Bigg),\end{aligned} \tag{4.29}$$

$$\begin{aligned}\bar{\theta}_i' = \varepsilon\Bigg(\delta_i & - \frac{r_{1i} f\cos\bar{\theta}_i}{2\bar{A}_i\eta_{0i}} \pm 4\frac{a\tan^2\alpha(r_{1i} - r_{2i})^2}{\eta_{0i}\pi}\arcsin\left(\frac{\Delta_1}{(r_{1i} - r_{2i})\bar{A}_i}\right) \\ & \pm 4\tan\alpha(a\Delta_1\tan\alpha - b)\frac{\sqrt{(r_{1i} - r_{2i})^2\bar{A}_i^2 - \Delta_1^2}}{\eta_{0i}\bar{A}_i^2\pi}\Bigg).\end{aligned} \tag{4.30}$$

For more details regarding the explicit integration of the nonlinear piecewise terms of the tuned wedge damper force, the reader is referred to appendix A. Equations (4.29) and (4.30) contain plus-or-minus signs as well as minus-or-plus signs. The upper signs correspond to the first modal coordinate, whereas the lower signs correspond to the second one. The equations above are only applicable if the relative amplitude is so large that the outer segments influence the dynamics of the system, i.e. $A_{\text{rel}} > \Delta_1 + r\sin\alpha$. Additionally, the existence conditions placed by the arcsin-term and the square root must be fulfilled. To obtain the stationary solution, the right hand sides of Eqs. (4.29) and (4.30) are equated to zero, i.e. $\bar{A}_i' = 0$ and $\bar{\theta}_i' = 0$. Solving these equations yields the system's response in the resonance regime. If the behavior at the resonance is sought, the phase difference $\bar{\theta}_i = -\pi/2$ is additionally plugged into the resulting equations, c.f. chapter 3. Thus, from Eq. (4.29) a relationship between the maximal amplitude and the excitation force is obtained. This relation is given by

$$\bar{A}_{i,\max} = \frac{1}{c\Theta_1}\left(\pi f r_{1i} + \Theta_2 + \sqrt{\pi^2 f^2 r_{1i}^2 + 2\Theta_2\pi r_{1i} f + cd\Theta_3 + d^2\Theta_4}\right) \tag{4.31}$$

with

$$\begin{aligned}\Theta_1 &= \mp 16(\tan^2\alpha + 1)\tan\alpha|r_{1i} - r_{2i}|(r_{1i} - r_{2i}),\\ \Theta_2 &= 8(\tan^2\alpha + 1)|r_{1i} - r_{2i}|(2c\Delta_1\tan\alpha - d),\\ \Theta_3 &= -256\Delta_1(r_{1i} - r_{2i})^2\tan\alpha(\tan^2\alpha + 1),\\ \Theta_4 &= 64(r_{1i} - r_{2i})^2(\tan^2\alpha + 1)^2.\end{aligned}$$

In Eq. (4.31) the value of the maximum amplitude is mainly determined by the linear terms in f. This confirms the robustness of the damper, i.e. the scalability of the frequency response function derived from Fig. 4.11. Additionally, the amplitude is inversely proportional to c, i.e. inversely proportional to the friction coefficient μ. Therefore, an increase in the friction coefficient leads to lower maximum amplitudes provided the parameters are chosen within the validity range of the asymptotic solution. The backbone curve of the system, the maximum amplitude dependent on the resonance frequency, is obtained in an implicit manner by an analogous approach with Eq. (4.30). This implicit equation is given by

$$\begin{aligned} 0 = \frac{\eta_{0i} - \eta}{\varepsilon} &\pm 4\frac{a \tan^2\alpha (r_{1i} - r_{2i})^2}{\eta_{0i}\pi} \arcsin\left(\frac{\Delta_1}{(r_{1i} - r_{2i})\bar{A}_{i,\max}}\right) \\ &\pm 4\tan\alpha (a\Delta_1 \tan\alpha - b)\frac{\sqrt{(r_{1i} - r_{2i})^2 \bar{A}_{i,\max}^2 - \Delta_1^2}}{\eta_{0i}\bar{A}_{i,\max}^2 \pi} . \end{aligned} \tag{4.32}$$

The analytical solution for the response function is verified by the numerical solution. To this end, an implicit equation for the response is derived by considering the stationary solution and eliminating θ_i. This implicit equation is then expressed in the original coordinates. A comparison for the standard parameters and $\varepsilon = 0.01$ is presented in Fig. 4.17. The solution lies within the acceptable range of the asymptotic method. Furthermore, the backbone curve accurately describes the maximal amplitude of the system. In contrast to the wedge damper, the backbone curve bends towards lower frequencies for decreasing amplitudes.

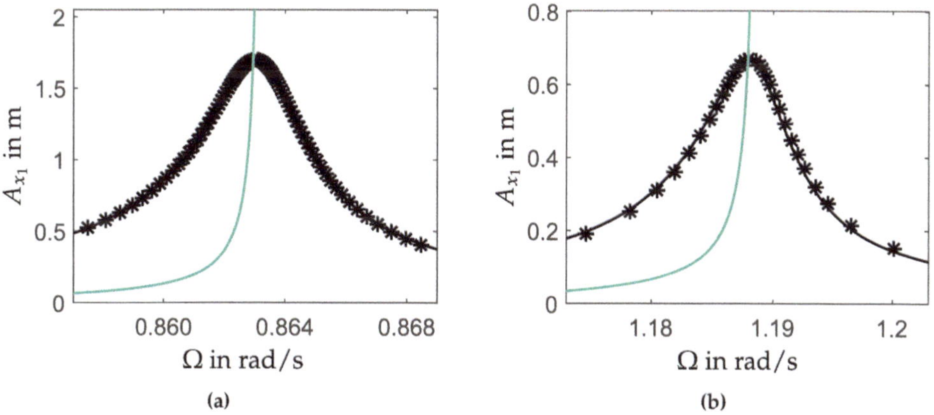

Figure 4.17: Comparison of the analytical (——) and the numerical solution ($*$), as well as the backbone curve (——). (a) $\Omega \approx \eta_{01}$. (b) $\Omega \approx \eta_{02}$.

4.4 Tuned Wedge Damper Optimization

To obtain an optimal damper response, key relationships derived from the analytical investigations are used as a starting point. First, the optimization function, as well as the optimization parameters, are defined from the analytical investigations. Second, optimizations with the analytical functions are carried out. Third, the scope of the analytical optimization is expanded by considering numerical methods. For the latter, the result of the analytical optimization is used as a starting point. Finally, the frequency responses of the optimized systems as well as their stability are considered.

The availability of an analytical solution allows the damper's targeted optimization. The numerical investigations show that the amplitude of the first eigenfrequency is dominant for the chosen parameter ranges. Therefore, a reduction of this amplitude optimizes the system's frequency response function. The prestress displacement $\Delta\ell$ is set to zero to favor the decoupling of absorption and damping. Furthermore, a variation of the friction coefficient is not considered. The limitations of this parameter are set by realizability conditions, life span requirements, or both. Once the limitations are set, the friction coefficient is chosen as high as possible. The Δ_1-parameter is set as large as necessary to ensure absorption. In practical applications, this is dependent on the natural occurring damping in the system. Since this again is a practical restriction on the system a variation of this parameter is not taken into account. The remaining parameters c_3 and α are considered for the optimization. The optimization function $g_{\text{opt}}(c_3, \alpha)$ is formulated and given by

$$g_{\text{opt}}(c_3, \alpha) = A_{x_1,\max}(c_3, \alpha) = |r_{11}|\bar{A}_{1,\max}(c_3, \alpha) \,. \tag{4.33}$$

The result of Eq. (4.33) is plotted in Fig. 4.18 for separate optimizations of c_3 and α. The optimizations are considered for $c_3 \in [0, 0.1]\,\text{N/m}$ and $\alpha \in [0°, 45°]$ since larger values compromise the validity range of the asymptotic solution. These figures reflect the behavior of the maximum amplitude observed in Figs. 4.5 and 4.6. For the standard parameters, the optimal values for the stiffness and angle are respectively $c_{3,\text{ana,opt}} = 0.059\,\text{N/m}$ and $\alpha_{\text{ana,opt}} = 40.12°$. These optimizations yield respectively maximal amplitudes of $A_{x_1,\max}(c_3) = 0.799\,\text{m}$ and $A_{x_1,\max}(\alpha) = 1.167\,\text{m}$, which are both lower than that of the standard parameters by respectively 52.96 % and 31.29 %. The separate optimizations show that the amplitude reduction has a higher sensitivity to the tertiary stiffness than to the angle of the outer segments.

The true potential of the analytical solution lies in the cost-effective multi-parameter optimization of the system's maximal amplitude. The joint optimization yields the

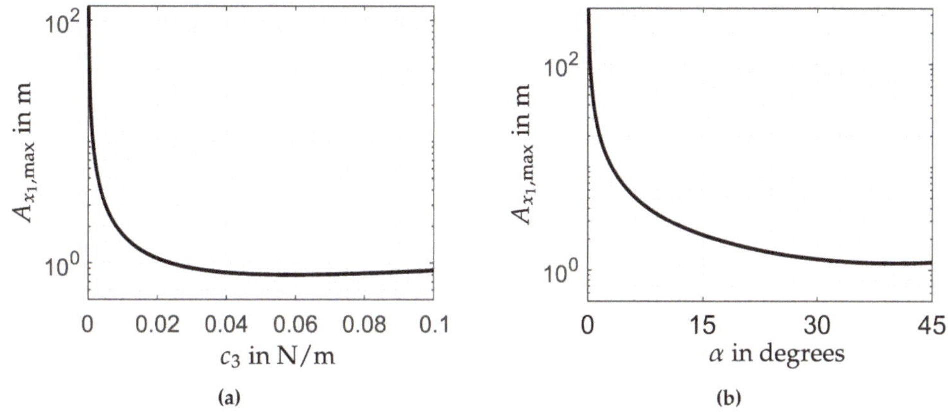

Figure 4.18: (a) Influence of the stiffness c_3 on A_{max}. (b) Influence of the wedge angle α on A_{max}.

parameters $c_3 = 0.1\,\mathrm{N/m}$ and $\alpha = 10.36°$. The maximum amplitude results in $A_{x_1,\mathrm{max}}(c_3, \alpha) = 0.547\,\mathrm{m}$ which means a 67.79 % amplitude reduction. It is noted that the optimized value for c_3 lies at the limit of the considered range. Therefore, further amplitude reduction potential is expected with an increase in this parameter. However, this is not within the scope of the asymptotic method.

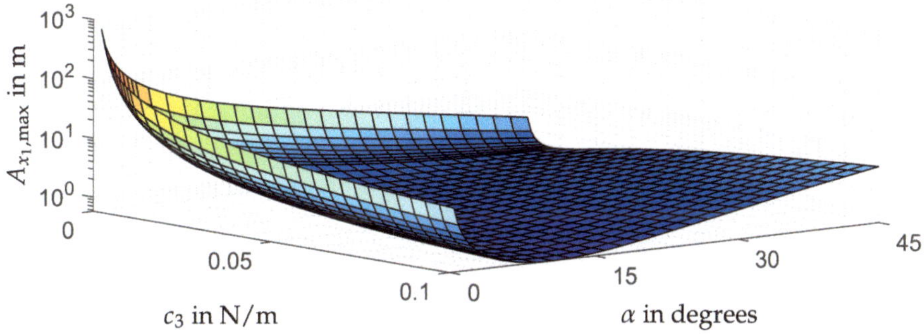

Figure 4.19: Influence of the stiffness c_3 and the wedge angle α on A_{max}.

Although the analytical optimization reaches its validity limits, it constitutes a starting point for further numerical optimizations. Analogous to Eq. (4.33), the maximum amplitude of the frequency response function is minimized, however, for each parameter combination the frequency response function is calculated with the harmonic balance method described in section 2.1. The computational cost is much higher than the evaluation Eq. (4.33) since the amplitudes for the frequency range $\Omega \in [0.5, 1.5]\,\mathrm{rad/s}$ are calculated instead of directly calculating the maximum amplitude. The parameters $(c_3, \alpha) = (0.1\,\mathrm{N/m}, 10.36°)$ are used to initialize a numerical optimization. The optimization is carried out with the *MATLAB* function `fminsearch`. This function implements a Nelder-Mead-Algorithm, which is a gradient-free optimization algorithm. This is essential since the problem is highly nonlinear and therefore gradient-based methods are costly. Furthermore, the accuracy of the gradient is not guaranteed. For more details on the Nelder-Mead-Algorithm, the reader is referred to [9].

Figure 4.20 shows the frequency response functions of the standard parameters, the analytical optimization, and two iterations of the numerical optimization. In addition, the stability and instability ranges are respectively marked with a solid and dashed line. In Fig. 4.20 the algorithm's 12th and 50th iterations are selected. The 50th iteration corresponds to the algorithm's final iteration since changes in the parameter are under the selected tolerance, namely 1e-3. The iterations result in the parameters

$$c_{3,\mathrm{opt,12th}} = 1.606\,\mathrm{N/m}\,, \qquad \alpha_{\mathrm{opt,12th}} = 1.55^\circ\,,$$
$$c_{3,\mathrm{opt,50th}} = 27.621\,\mathrm{N/m}\,, \qquad \alpha_{\mathrm{opt,50th}} = 1.09^\circ\,.$$

As is noted from the parameters, with increasing iteration the stiffness rises and the wedge angle declines. Low wedge angles are advantageous for the smooth transition between the individual segments. Furthermore, the damping is introduced mainly via the third spring and not via the wedge angle. Although the peaks of the 50th optimization iteration are close to another, all responses show their maximum amplitude at the first resonance peak. The 12th and 50th iterations of numerical optimizations have maximum amplitude values of 0.213 m and 0.062 m respectively. Compared to the system with the standard parameters, these values correspond to amplitude reduction of 87.47 % and 96.38 %. The higher iterations achieve a larger amplitude reduction, however, this comes at a cost. In certain frequency ranges, the stability of the periodic solution is lost.

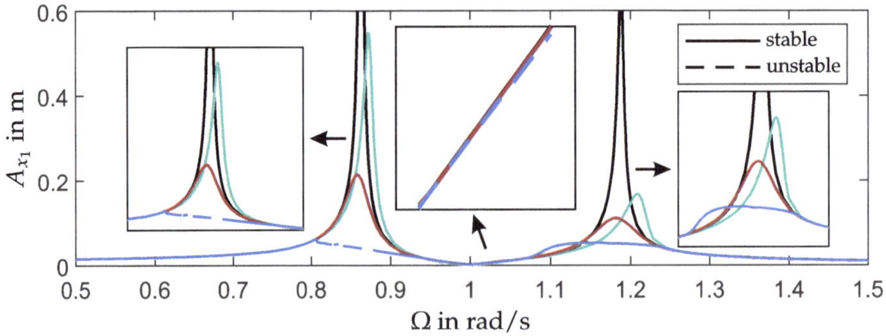

Figure 4.20: Frequency response functions of the standard parameters (——), the analytical optimization (——), the 12th iteration of the numerical optimization (——), and the 50th iteration of numerical optimization (——).

This occurs in the vicinity of the resonance regimes as well as in the absorption range. Although the oscillations are small and absorption is observed, the relative amplitudes in the absorption range are larger than Δ_1. Subsequently, the strong nonlinearities introduced by the outer segments lead to unstable solutions. Due to these stability losses, the 50th iteration is not a reliable solution. The 12th iteration shows the lowest amplitude without the stability loss throughout the whole frequency range.

In order to assess the disturbance of the frequency response due to the unstable ranges, sweep simulations of the tuned wedge damper are carried out. In comparison to a simple harmonic excitation, the sweep simulation offers the advantage that it evaluates multiple frequencies with one simulation. Additionally, by calculating the envelope of the sweep response an approximation of the system's amplitude response is derived. The results are not identical but comparable to the solution of the harmonic balance. Furthermore, since a forward-time simulation is carried out, only stable solutions are observed in the results. Therefore, variations are expected in unstable regions. To implement such an excitation, the harmonic force in Eq. (4.1) is replaced by the sweep excitation $F_{\text{sweep}}(t) = F \sin(\varphi_{\text{sweep}}(t))$ with

$$\varphi_{\text{sweep}}(t) = \int_0^t \Omega_{\text{sweep}}(\tau)\mathrm{d}\tau \,, \quad \Omega_{\text{sweep}}(t) = \frac{\Omega_1 - \Omega_0}{t_1 - t_0}(t - t_0) + \Omega_0$$
$$F = 0.01\,N,\ \Omega_0 = 0\,\mathrm{rad/s},\ \Omega_1 = 2\,\mathrm{rad/s},\ t_0 = 0\,\mathrm{s}\,,\ t_1 = 10^5\,\mathrm{s}\,.$$

The results of the sweep excitations are plotted in Fig. 4.21. The response of the system with the standard parameters, the analytical optimization, and the 12th iteration of the numerical optimization show a similar response as in Fig. 4.20. Due to the instability ranges substantial differences are noted between the sweep response of the 50th iteration

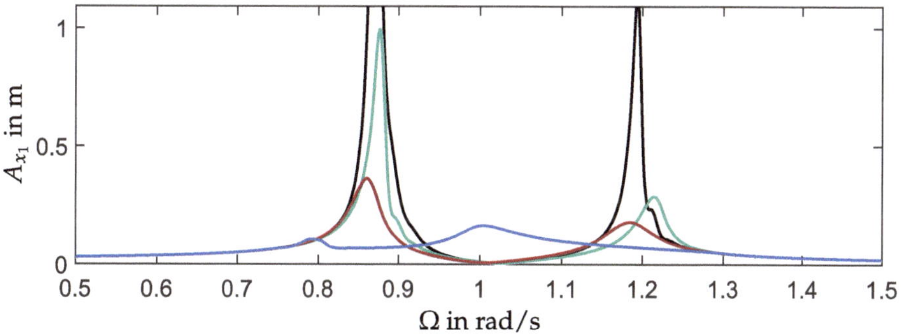

Figure 4.21: Envelopes of the sweep response for the standard parameters (——) the analytical optimization (——), the 12th optimization iteration (——), and the 50th optimization iteration (——).

and the harmonic balance solution, c.f Fig. 4.20. The absorption range and with it the decoupling of damping and absorption are completely lost. Furthermore, the vibration reduction observed in the harmonic balance solution is not observed. Instead, the maximum amplitude of 0.161 m is noted. However, the maximum amplitudes of the 50th iteration are still lower than those of the 12th iteration. If only the minimization of the maximum amplitude is sought, a numerical optimization, which considers a frequency response function obtained via sweep calculations, is a more sensible approach. However, since the damper's goal is partly to retain the advantages of absorption, the 12th iteration is considered the optimal response of the tuned wedge damper for the standard parameters.

4.5 Damper Assessment

The tuned wedge damper provides an advantageous device for vibration reduction. It reduces the system's vibrations while maintaining an absorption frequency. A separation of absorption and damping is achieved by placing the classical vibration absorber between specially designed contact surfaces. No damping is introduced when the oscillations are small and in the Δ_1-range. If the oscillations are larger, the outer segments come into play. Energy is damped via dry friction of the contact between the secondary mass and the outer segments. Amplitude reductions are found with increasing friction coefficients. An effective amplitude reduction potential is obtained through the simultaneous optimization of the tertiary stiffness c_3 and the wedge angle α via analytical and numerical methods.

The decoupling achieved by this damper is advantageous especially for machines during run-up procedures since one or more eigenfrequencies usually have to be overcome. In comparison to conventional friction dampers with a constant force, the tuned wedge damper also provides a robust solution. This is especially useful in overload scenarios.

Limits on the amplitude reduction are placed by the realizable friction coefficients. High friction coefficients lead to higher wear which affects the service life of the damper. The exceeded increase of the friction coefficient, therefore, inhibits the damper's functionality in the long-term. The choice of the friction coefficient results from the counterbalancing between amplitude reduction and life span requirements. Furthermore, the design freedom of the friction coefficient is decreased in scenarios where additional constraints are taken into account, e.g. weight, corrosion, or production constraints. In these scenarios, alternatives to the tuned wedge damper are necessary. An active design of the tuned wedge damper would amplify its intrinsic targeted dissipation.

5 Experimental Investigations

In chapters 3 and 4 theoretical insight was obtained on the friction-based dampers. This insight is tested via experimental investigations in this chapter. The goal is therefore to verify the qualitative aspects of the dampers in more realistic conditions. A quantitative reproduction of the results is not pursued. Furthermore, this chapter presents the first proof of concept for the implementation of these friction-based dampers. This chapter is structured as follows. First, the experimental setup is described in section 5.1. Experimental investigations into the wedge damper and tuned wedge damper are subsequently presented in sections 5.2 and 5.3. The chapter is closed with the experimental conclusions in section 5.4.

5.1 Experimental Setup

To describe the experimental setup, first, the general arrangement of the experiments is presented. Subsequently, its components are described in further detail. Second, the mechanical components and the design of the friction damper are considered. Third, the inputs and outputs of the system, namely the shaker and sensors, are presented. Finally, the software, as well as the post-processing is described.

The test bench is presented and highlighted in Fig. 5.1 and is composed of the vibration isolation table, the system prototype, the power amplifier, the PULSE front-end driver, and a laptop with the PULSE software. The laptop determines the measurement parameters, as well as the input signals. The interface between the laptop and the shaker/sensors is the PULSE front-end driver. This component implements the measurement parameters defined by the laptop, e.g. sampling rate. The input signal of the shaker is generated and sent to the power amplifier. This signal is amplified and transferred to the shaker. The shaker excites the system and causes vibrations which are measured by accelerometers. A force sensor is placed to measure the force that acts on the system. The sensors send the signals to the PULSE front-end driver, which converts them into processable data for the laptop.

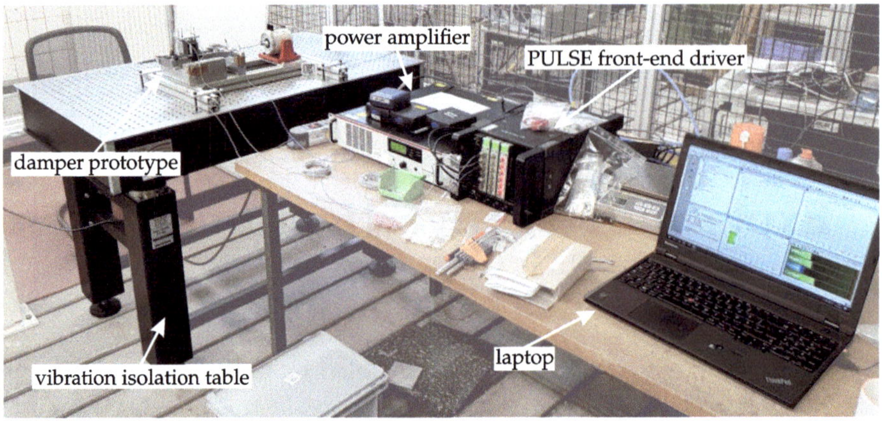

Figure 5.1: The experimental setup for the investigations.

The prototype is 3D printed with an Ultimaker 5 and includes the main system and the friction-based damper. Both subsystems are highlighted in Fig. 5.2 and printed with polylactide fibers. The main system is highlighted in dark gray, whereas the friction damper is accentuated in light gray. Since the wedge damper is a special case of the tuned wedge damper, the experimental setup is described for the latter. Furthermore, the prototype is a realization of the schematic systems described in chapters 3 and 4. However, this is only partially true. The largest difference to the schematic systems is the additional damping caused by friction in the guides and joints and material damping. Another discrepancy lies in the elasticity of the experimental parts which are in the minimal models assumed rigid.

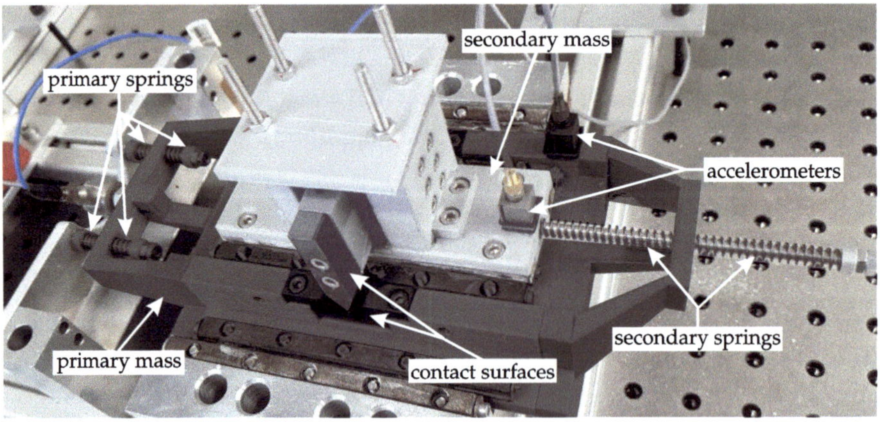

Figure 5.2: The prototype of the main system and the friction-based damper.

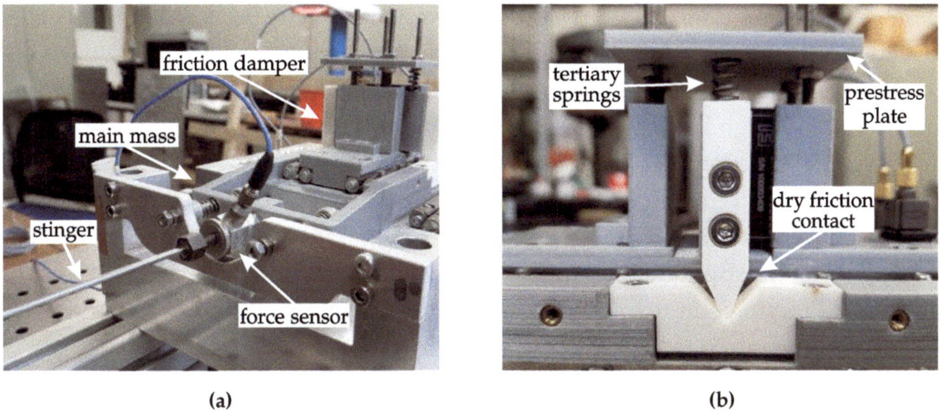

Figure 5.3: (a) The shaker-system coupling with a stinger. (b) The design of the friction contact as well as the pressing arrangement.

The main system is composed of the main mass and the main spring. The main mass is constructed with inner steel cylinders to achieve the desired weight of approximately 1 kg. Additionally, it is excited by a shaker-stinger arrangement, see Fig. 5.3a. By bending, the stinger compensates for any manufacturing inaccuracies and ensures a horizontal force. Furthermore, the main mass is mounted onto the environment with ERO type R linear guides. This arrangement allows a movement in the horizontal direction. The main spring is realized by four VD143Q Gutekunst Feder compression springs, that are mounted to realize a symmetrical spring force in both movement directions.

The friction damper includes the secondary mass and the secondary and tertiary springs. The secondary mass is entirely 3D printed and weighs approximately 300 g. It is connected to the main mass in three ways. It is mounted on top of the main mass with ERO type R linear guides so that both masses move parallel to each other. It is also connected via the symmetrically mounted secondary springs and via the dry friction contact. The design of the contact components is modular to vary the contact properties and is presented in Fig. 5.3b. They are also designed with smooth transitions to avoid impacts. The smallest roundings in the contact surfaces have a radius of 0.3 mm. To investigate different friction coefficients the contact components are printed in three different materials: IGUS IGLIDUR I180 (I180), polylactide (PLA), and thermoplastic polyurethane (TPU). The IGUS-polymer is designed for low friction and high wear resistance. PLA is a polymer commonly used printing material with little compliance. Lastly, TPU is a highly compliant thermoplastic. For a homogeneous material pairing, friction is lowest in IGUS, medium in PLA, and highest in TPU. Figure 5.3b also presents the design mechanism by which the tertiary springs press the contact surfaces together. The linear guide allows the contact component of the friction damper a vertical movement, whereas the contact component of the main system is fixed. Via a movement of the prestress plate above the tertiary spring the prestress of the system is varied. It is noted that this design is not symmetrical and presents an additional degree of freedom. However, an interaction of this degree of freedom with the horizontal movement of the damper is not observed. Additionally, the force of the friction dampers is reduced, since a relative movement in this arrangement causes half of the spring displacement. Thus, for a vanishing prestress the theoretical force of the damper prototype is by a factor 2 smaller than the force in the schematic dampers in Figs. 3.1 and 4.1. In contrast, the effect of the prestress is unaffected by the design.

The experimental setup contains one shaker, one force sensor, and two accelerometers. The shaker excites the system and is a LDS V201 Shaker with permanent magnets from Brüel & Kjær. The force sensor is a PCB 208C03 force sensor and is placed between the stinger and the main system. This placement ensures that the force that acts on the main system is measured. This way the dynamics of the shaker and stinger are excluded from the measurements. It is noted that the force drop effect is observed while passing through the resonance regimes. In these regimes, the oscillations are large enough, that the magnetic field in the shaker becomes nonlinear and the force acting on the system drops [171]. An active control, to avoid this effect, was not pursued, since the effect does not affect the qualitative influence of the parameters on the damper's behavior. The accelerometers are piezoelectric with a seismic mass and also from Brüel & Kjær of type 4507 B004. They are placed on the main and secondary mass and measure the acceleration. The measured signals are afterwards post-processed.

The PULSE software *LabShop* is used to determine the measurement parameters. First, the input signal for the shaker is defined as a sweep-excitation. The frequencies from 6 Hz to 40 Hz are passed through in 400 s. The system is excited and the resulting vibrations are measured. A modified low-pass third order Butterworth-filter with a cut-off frequency of 400 Hz is applied internally by the *LapShop*-Software. The sampling frequency is set to 1024 Hz which is above the necessary frequency to fulfill the Nyquist-Shannon sampling criterion and ensures an accurate signal reconstruction. The three measured signals are exported and post-processed in *MATLAB*. A Butterworth high-pass filter of fifth-order with a cut-off frequency of 4 Hz is applied to the measured signals. This cuts off the lower frequency components and enables the subsequent integration. The velocity and displacement signals of the primary and secondary mass are calculated with a trapezoidal numerical integration. Since a sweep-excitation is applied, the envelope of the measurement yields an approximation of the system's amplitude response function. It is noted that only stable solutions are present in the measurement. Therefore, the response is only a valid approximation if the frequency ranges with multiple solutions are small. In contrast, if the ranges with multiple solutions are pronounced, jumps in the measurement are expected. To obtain accurate measurements, five envelopes are averaged. If the amplification factor is sought, an H_2 estimator is used since it presents an upper bound for the transfer function. It is also more accurate in the resonance regimes than for example the H_1 estimator. A disadvantage of this estimator is that it yields a linear approximation of a nonlinear system. Therefore, it is only applied in the variation of the excitation force and for solutions in which no jumps are noted. The H_2 estimator is calculated as

$$H_2(\omega) = \left| \frac{S_{\tilde{X}\tilde{X}}(i\omega)}{S_{\tilde{X}\tilde{F}}(i\omega)} \right| . \tag{5.1}$$

where $S_{\tilde{X}\tilde{X}}(i\omega)$ is the autospectrum of the output and $S_{\tilde{X}\tilde{F}}(i\omega)$ represents the cross-spectrum between output and input. Additionally, i is in this case the imaginary unit and ω the considered angular frequency. These terms are calculated from the post-processed data with

$$S_{\tilde{X}\tilde{X}}(i\omega) = \frac{1}{N}\sum_{j=1}^{N} \overline{X}_j(i\omega)X_j(i\omega) \quad \text{and} \quad S_{\tilde{X}\tilde{F}}(i\omega) = \frac{1}{N}\sum_{j=1}^{N} \overline{X}_j(i\omega)F_j(i\omega) . \tag{5.2}$$

$F_j(i\omega)$ and $X_j(i\omega)$ represent the Fourier transformations of the input force $F(t)$ and the output $x_1(t)$ of the j-th measurement. The bar $(\overline{\ \ })$ denotes the complex conjugate. To reduce leakage, the post-processed signals are multiplied by a Hanning window.

5.2 Wedge Damper Experiments

The experimental investigations consider the effect of parameter variations on the response of the system. The setup for the wedge damper removes the spring between the primary and secondary mass. The spring which presses the contact surface together is thus the secondary spring. A D205D spring from Gutekunst Ferder is chosen. For the contact surfaces a I180-PLA material pairing is selected, due to its low friction coefficient. Since the force drop effect is observed, a single force amplitude is not available. Instead, the excitation force is characterized by the root-mean-square (rms) value of the measured oscillations. The standard parameters for the wedge damper setup are given by

$$m_1 = 1.057\,\text{kg},\ m_2 = 0.305\,\text{kg},\ c_1 = 15.928\,\text{N/mm},\ c_2 = 24.232\,\text{N/mm},$$
$$\alpha = 30^\circ,\ \mu = \mu_{\text{I180-PLA}},\ \Delta\ell = 0\,\text{mm},\ F = 4.072\,\text{Nrms} .$$

The first experimental parameter variation changes the wedge angle α of the lower contact surface and is presented in Fig. 5.4. The figure confirms the qualitative effect of the angle change in the wedge damper. The effective secondary stiffness changes as in a conventional tuned mass damper. With an increase in the angle, both peaks are shifted towards higher frequencies. Moreover, as the peak of the first eigenfrequency increases, the second decreases. This agrees with the results of Fig. 3.5.

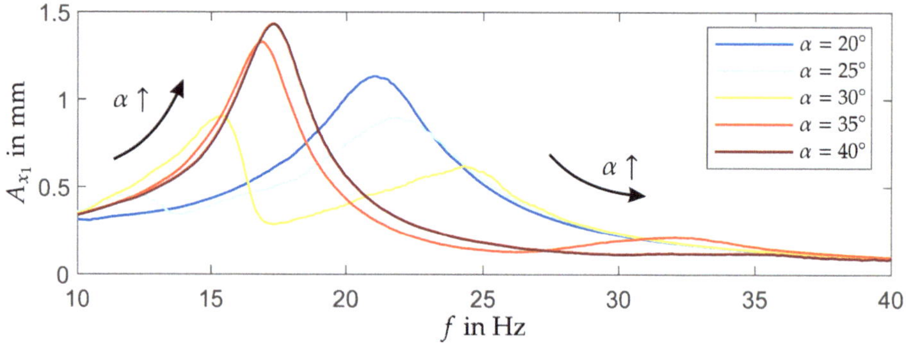

Figure 5.4: Wedge Damper experimental variation of the wedge angle α.

Figure 5.5 shows the results of the variation of the friction coefficient. The friction coefficient is varied by considering different material pairings. Since the friction coefficient is not in the focus of this work, a detailed investigation into its exact value is not considered. Instead, the curves are arranged by their expected friction value. Although scattering in the friction coefficient is expected, the curves show the increase in damping due to the increase in friction. As the friction coefficient increases, the resonance peaks merge, and only one peak is observed. Deviations from this general

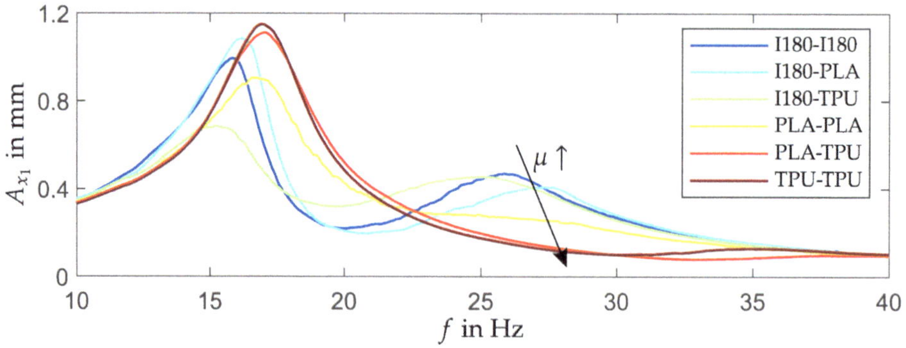

Figure 5.5: Wedge Damper experimental variation of the friction coefficient μ.

tendency are observed in the curves with a TPU pairing. This is in part due to the additional compliance introduced by the elasticity of this material.

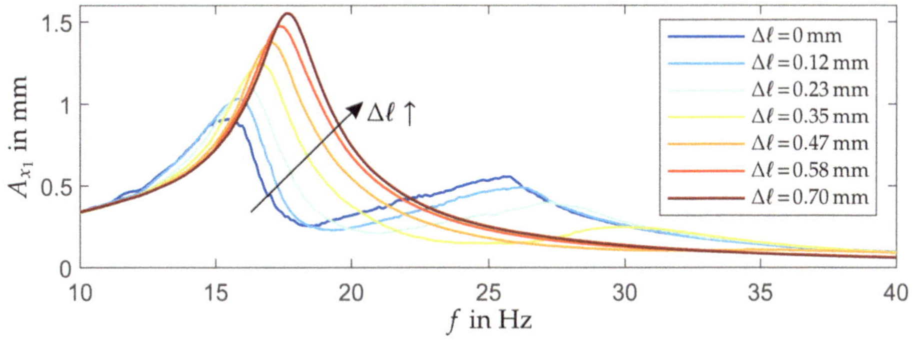

Figure 5.6: Wedge Damper experimental variation of the prestress displacement $\Delta\ell$.

The variation of the prestress displacement $\Delta\ell$ is presented in Fig. 5.6. This experimental consideration also coincides with the numerical results in section 3.2. As the prestress increases the amplitude of the first peak rises and the secondary peak vanishes. The system tends to a one-degree of freedom system where only sticking is observed. Thus, absorption is lost. An increase of the first resonance peak for prestresses higher than 0.70 mm is not observed. This increase is prevented by the additional material damping of the system. In contrast to the numerical simulations, jumps in the amplitude response are not observed. This is at least partially due to the inherent system damping since it reduces the system's maximal amplitudes.

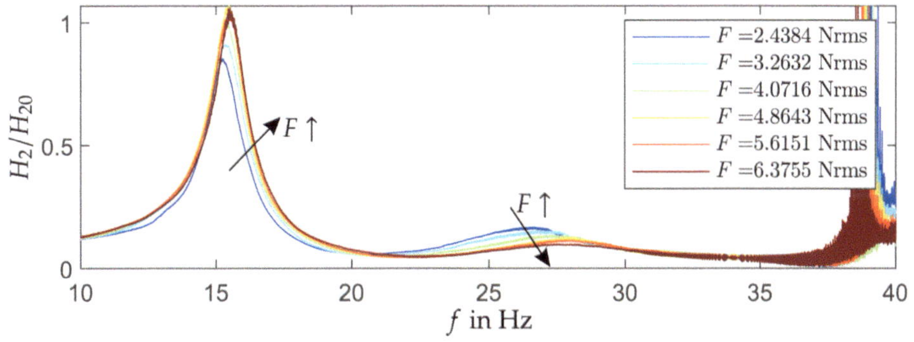

Figure 5.7: Wedge Damper experimental variation of the excitation force F.

Figure 5.7 presents the investigations of the excitation force F and its effect on the normalized H_2 estimator. The estimator is normalized by the maximum H_2 value for the standard parameters. The curves lie approximately over each other. Only differences at low excitation forces and at the second resonance frequency are noted. At low excitations, the dry friction in the guides and joints substantially contributes to vibration reduction. However, at higher excitation forces this is not the case, since the reduction in amplitude is due to the friction damper dissipation. The estimator curves are approximately equal which implies the scalability of the response function and the robustness of the damper. The noise at the end of the frequency range occurs due to the noise in the acceleration $x_1(t)$ since the H_2 estimator is especially sensible to the output noise.

5.3 Tuned Wedge Damper Experiments

As with the experimental investigations of the wedge damper, selected parameter variations are considered. The setup for the tuned wedge damper is as described in section 5.1. For the secondary and tertiary spring, a D173CJ and a VD097 from Gutekunst Feder are respectively selected. The standard material pairing is a PLA-PLA combination. The standard parameters for the tuned wedge damper setup result in

$$m_1 = 1.065\,\mathrm{kg},\ m_2 = 0.313\,\mathrm{kg},\ c_1 = 15.928\,\mathrm{N/mm},\ c_2 = 5.948\,\mathrm{N/mm},$$
$$c_3 = 2.004\,\mathrm{N/mm},\ \alpha = 45°,\ \mu = \mu_{\mathrm{PLA-PLA}},\ \Delta\ell = 0\,\mathrm{mm},\ \Delta_1 = 0.5\,\mathrm{mm},$$
$$r = 0.5\,\mathrm{mm},\ F = 3.151\,\mathrm{Nrms}.$$

The first experimental parameter variation considers the variation of the tertiary stiffness and is presented in Fig. 5.8. The qualitative behavior on both resonance frequencies is verified. An optimal stiffness exists and yields minimal amplitudes in the first resonance peak. Furthermore, the amplitudes of the secondary peak sink with rising stiffness values. This confirms the behavior in Fig. 4.4. Additionally, the absorption frequency remains largely unaffected. Minimal changes are observed in its position however not

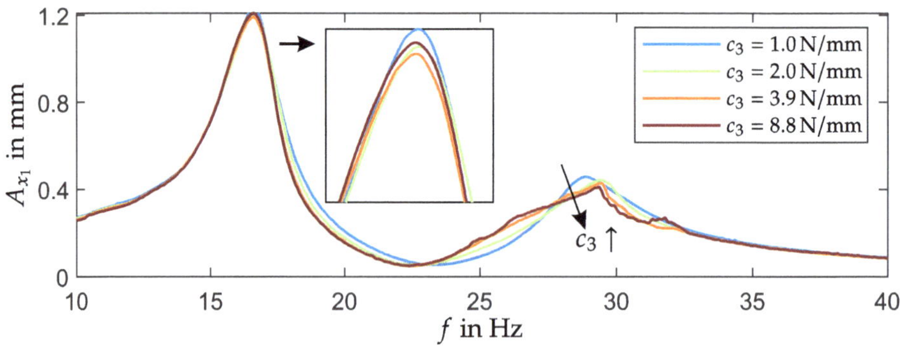

Figure 5.8: Tuned Wedge Damper experimental variation of the tertiary stiffness c_3.

in the amplitude. The changes in the position are due to imperfections in the assembly of the contact surfaces.

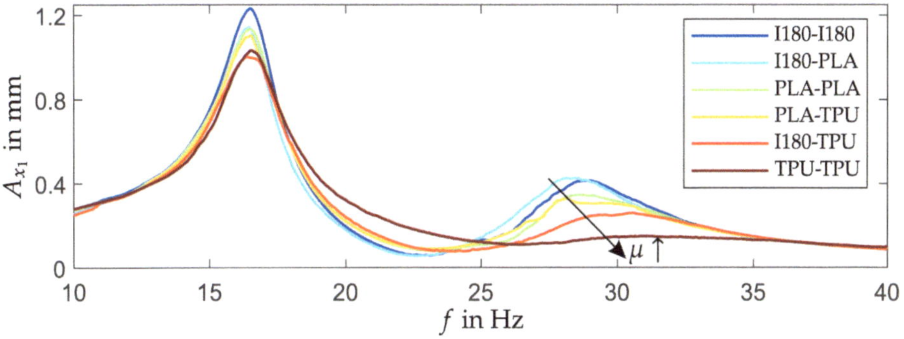

Figure 5.9: Tuned Wedge Damper experimental variation of the friction coefficient via material changes.

Figure 5.9 presents the investigations into the friction coefficient via the changes in material pairings described above. The decoupling between vibration damping and vibration absorption is observed. Both peaks show a substantial reduction and the absorption frequency remains largely unaffected. Due to the high elasticity of the TPU material, deviations from the general tendency are again observed in some experiments. However, the decoupling effect is still observed. This is evident when comparing these results with Fig. 5.5. In contrast to the wedge damper, the maximum amplitudes are reduced and a merging of the two resonance peaks into one is not observed.

The prestress displacement is varied in Fig. 5.10. In contrast to the results of the wedge damper, the amplitudes do not rise monotonically with increasing prestress values. A shift to a one-degree of freedom system is observed, and an optimal prestress value

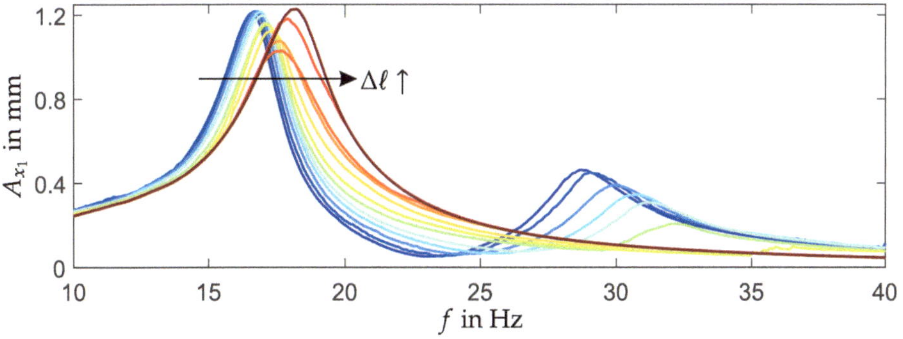

Figure 5.10: Tuned Wedge Damper experimental variation of the prestress displacement $\Delta\ell$ from $\Delta\ell = 0\,\text{mm}$ (——) to $\Delta\ell = 4.9\,\text{mm}$ (——) in 0.23 mm steps.

for the first resonance peak is noted. This confirms the numerical results in Fig. 4.10. Furthermore, as the system transitions into complete sticking the second eigenfrequency vanishes.

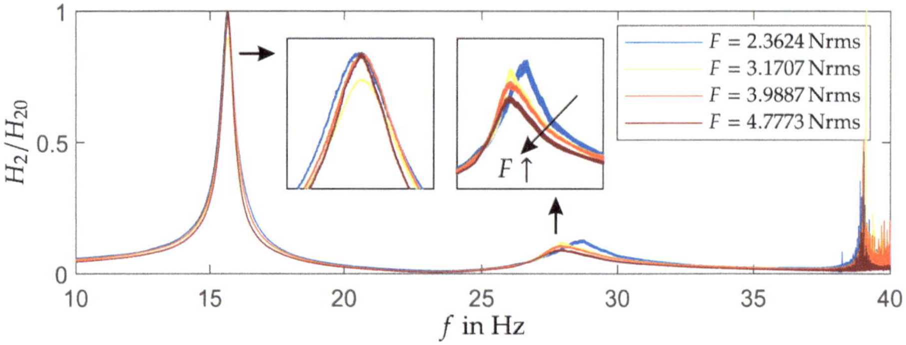

Figure 5.11: Tuned Wedge Damper experimental variation of the excitation force F.

The last experimental study handles the variation of the excitation force and is presented in Fig. 5.11. The estimator H_2 is again normalized with the value H_{20}, which corresponds to the maximum H_2-value of the standard parameters. The curves lie nearly over each other. This implies the approximate scalability of the estimator H_2. Differences are noted at the second eigenfrequency. The H_2 values are lower for higher excitation forces, which implies higher damping with increasing amplitudes. This is caused by the increased influence of the outer dissipative segments compared to the constant influence of the dissipation-free segment.

5.4 Experimental Conclusions

The investigations above offer the first proof of concept for the realization of these friction-based dampers. The theoretical behavior of the damper is modified by the real conditions of the experiment which introduce additional damping and elasticities. Some aspects of the dampers' behavior were confirmed.

Quasiperiodic and chaotic solutions were not observed. These types of solutions appear as irregularities in measurements. However, no such deviations at the expected ranges were observed. Jumps in the measurements, that imply multiple solutions, were also not detected. A possible explanation for the vanishing of such solutions is the increased damping provided by the experimental conditions. The vanishing of multiple solution ranges due to increased damping, for example, is also observed with the multiple solution ranges of systems with cubic stiffness.

The wedge damper's behavior as a pseudo-viscous damper is confirmed. A change in the wedge angle α leads to a change in the effective stiffness and damping, whereas the change in the friction coefficient affects the damper's effective damping. Additionally, the increase in amplitudes due to the prestress displacement is also confirmed. The quasi-viscous behavior of the wedge damper is also verified by the variations of the excitation force and its influence on the H_2 estimator.

The response of the tuned wedge damper is also validated via experiments. The existence of an optimal tertiary stiffness is observed, as well as the existence of an optimal prestress value for the first resonance peak. As with the wedge damper, the scalability of the response is also confirmed with the variation of the excitation force. Lastly, the decoupling of vibration damping and vibration absorption, which is the main advantage of the damper, is experimentally verified.

6 Active Tuned Wedge Damper

The investigated friction dampers offer effective vibration reduction alternatives in comparison to conventional dampers. However, in specific cases, the advantages these dampers provide are not sufficient. Such cases include exceedingly long service life requirements or cases where the material pairing is not freely chosen, but subject to constraints. Such cases especially those with low friction coefficients cannot ensure an adequate vibration reduction. An active consideration of the tuned wedge damper offers a vibration reduction solution with small friction coefficients. Additionally, in situations in which extremely low amplitudes are required, the amplitude reduction provided by the optimized tuned wedge damper may not be sufficient. The active tuned wedge damper offers a solution with more selective energy dissipation. This chapter focuses on three such active considerations. First, the base system for the control strategies and the framework for the simulations are presented in section 6.1. Second, a Skyhook Control is considered in section 6.2. This is a well-investigated control strategy for friction dampers and is taken as a reference point for the control strategies developed in this work. Third, in section 6.3 an adaptive control strategy with multiple actuation models is detailed. Section 6.4 focuses on a slow frequency-based control. The efficiencies of the control strategies with respect to the amplitude reduction are compared in section 6.5, whereas a consideration of the dissipated energy is made in section 6.6. The investigations into active friction dampers are concluded in section 6.7 with an assessment of the control strategies.

6.1 Base System for the Control Strategies

This section presents the general conditions for the investigations of the active system and is divided into two parts. First, the base system, on which the control strategies are applied, is presented. Additionally, the input parameters as well as their effect on the damper force are detailed. The second part presents the framework of the simulations that are used to investigate the control strategies. This includes the excitation types, the initial conditions, and the constant damper parameters.

The tuned wedge damper is chosen as a base system for the control strategies since the wedge damper is a special case of the former. Furthermore, it has the advantage of decoupling damping and absorption. The active tuned wedge damper is presented in Fig. 6.1. The main system is represented by the main mass m_1 and the main spring c_1. It is also excited by the force $F(t)$, which is highlighted in red. From the control point of view, this is a perturbation on the otherwise resting main system. It is the task of the active tuned wedge damper and its control strategy to counter this perturbation. To this end, the active wedge damper is attached to the main system. The friction damper is composed of the secondary mass m_2, the secondary spring c_2, the tertiary spring c_3 (with its prestress displacement $\Delta\ell$), and the contact surfaces. The last component is defined by the wedge angle α, the length of the horizontal segment $2\Delta_1$, and the radius of the transition segments r. The positions of the main and secondary mass are respectively described by x_1 and x_2. For a more detailed description of the underlying mechanical system, the reader is referred to chapter 4. The controlled parameters of the active tuned wedge damper are highlighted in green. Thus, the damper varies the wedge angle α of the outer segments and the prestress displacement $\Delta\ell$.

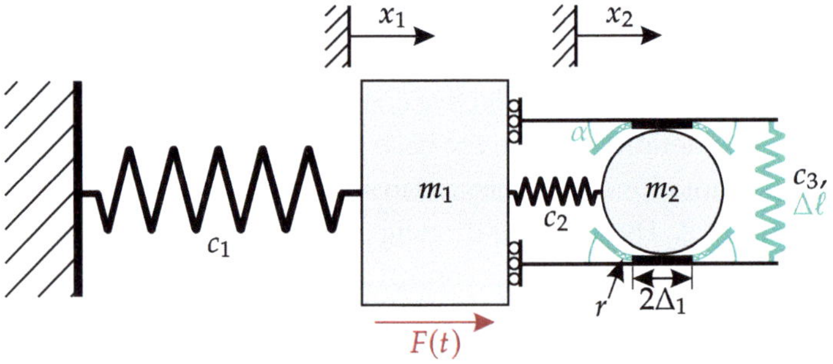

Figure 6.1: The schematic model of the active tuned wedge damper.

The qualitative influence of the input parameters α and $\Delta\ell$ on the damper force F_{TWD}, see Eq. (4.6), is presented in Fig. 6.2. An increase in both of these parameters leads to more damping. The difference lies in the way the damping is introduced. Higher values of the angle α lead to a damping increase only in the outer segments, whereas higher prestress values increase the damping over the whole oscillation. From a geometrical point of view, the wedge angle changes the opening angle of the outer force hysteresis, and the prestress displacement changes the width of the whole force hysteresis. Furthermore, their influence on the dissipated energy is also different. The higher angles increase the dissipated energy proportional to the square of the relative amplitude A_{rel}^2. In contrast, the prestress increment affects only the dissipated energy proportional to the relative amplitude A_{rel}, c.f. Eq. (4.17). The control strategies vary both of these parameters to introduce damping in a targeted manner with low friction coefficients.

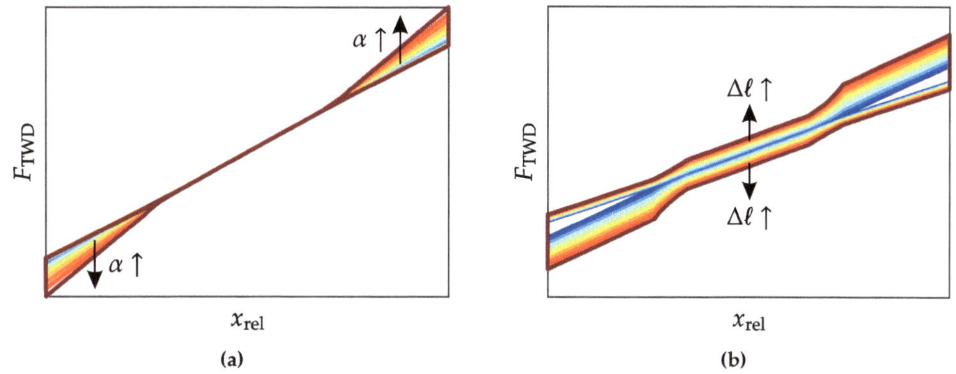

Figure 6.2: Effects of the input parameters on the tuned wedge damper force: (a) wedge angle α variation for $\Delta\ell = 0\,\mathrm{m}$ and (b) prestress displacement $\Delta\ell$ variation.

The following subsections compare the control strategies to their passive counterpart. Three excitation cases are considered: a rectangular pulse excitation, a sweep excitation, and an application scenario. The rectangular pulse gives insight into the transient behavior of the strategies, whereas the sweep excitation gives an approximation for the stationary behavior. The application scenario provides information about the suitability of the control strategies for more realistic applications. The sum of insights obtained via the excitation simulations yields the theoretical basis for an active implementation of the tuned wedge damper.

In the first case, the system is let go from nontrivial initial conditions, and the free behavior is investigated. Afterward, a force in the form of a rectangular pulse is applied. This excitation yields insights into the transient behavior of the control strategies. Furthermore, understanding into the strategies' reaction to harsh excitation is obtained. The excitation force in this case is given by

$$F_{\mathrm{rect}}(t) = \begin{cases} 0, & t_0 \leq t < t_1 \\ F_0, & t_1 \leq t < t_2 \\ 0, & t_2 \leq t < t_3 \end{cases}$$

with

$$F_0 = 0.01\,\mathrm{N},\; t_0 = 0\,\mathrm{s},\; t_1 = 200\,\mathrm{s},\; t_2 = 300\,\mathrm{s},\; t_3 = 325\,\mathrm{s}\,.$$

The second case applies a sweep excitation and focuses on the approximation of the system's stationary behavior. The system starts with zero initial conditions, and a frequency range including both structural resonances and the absorption frequency is slowly passed through. This yields an approximation of the control strategies' response at different excitation frequencies. The sweep excitation is described by

$$F_{\text{sweep}}(t) = F_0 \sin\big(\varphi_{\text{sweep}}(t)\big)$$

with

$$\varphi_{\text{sweep}}(t) = \int_0^t \Omega_{\text{sweep}}(\tau)\mathrm{d}\tau\,,\ \Omega_{\text{sweep}}(t) = \frac{\Omega_1 - \Omega_0}{t_1 - t_0}(t - t_0) + \Omega_0\,,$$
$$F_0 = 0.01\,\text{N},\ t_0 = 0\,\text{s}\,,\ t_1 = 10^5\,\text{s}\,,\ \Omega_0 = 0\,\text{rad/s},\ \Omega_1 = 2\,\text{rad/s}\,.$$

The third case considers an application scenario. First, a run-up to the system's operation frequency is performed starting from zero initial conditions. This is followed by a nominal operation range with a harmonic excitation. Subsequently, an overload section is simulated, in which the excitation amplitude rises tenfold. Afterward, the excitation is brought again to its original level in the recovery phase. Finally, the system is shut down in a controlled manner. In this phase, the excitation force and the excitation frequency linearly approach zero. The consideration of both the system run-up and run-down has the advantage that nonlinear phenomena in the system response are taken into account. The scenario force is given by

$$F_{\text{scen}}(t) = F_{\text{A,scen}}(t)\sin(\varphi_{\text{scen}}(t))\,.$$

The components $F_{\text{A,scen}}(t)$ and $\Omega_{\text{scen}}(t)$ are visualized in Fig. 6.3 and given by

$$F_{\text{A,scen}}(t) = \begin{cases} F_0, & t_0 \le t < t_2 \\ \dfrac{F_3 - F_2}{t_3 - t_2}(t - t_3) + F_3, & t_2 \le t < t_3 \\ \dfrac{F_4 - F_3}{t_4 - t_3}(t - t_4) + F_4, & t_3 \le t < t_4 \\ \dfrac{F_5 - F_4}{t_5 - t_4}(t - t_5) + F_5, & t_4 \le t < t_5 \end{cases},$$

$$\varphi_{\text{scen}}(t) = \int_0^t \Omega_{\text{scen}}(\tau)\mathrm{d}\tau\,,\ \Omega_{\text{scen}}(t) = \begin{cases} \dfrac{\Omega_1 - \Omega_0}{t_1 - t_0}(t - t_0) + \Omega_0, & t_0 \le t < t_1 \\ \Omega_1, & t_1 \le t < t_4 \\ \dfrac{\Omega_5 - \Omega_4}{t_5 - t_4}(t - t_5) + \Omega_5, & t_4 \le t < t_5 \end{cases},$$
$$F_0 = F_1 = F_2 = 0.01\,\text{N},\ ,\ F_3 = 10F_0,\ F_4 = F_0,\ F_5 = 0\,\text{N},$$
$$\Omega_0 = 0\,\text{rad/s},\ \Omega_1 = \Omega_2 = \Omega_3 = \Omega_4 = 1\,\text{rad/s},\ \Omega_5 = 0\,\text{rad/s},$$
$$t_0 = 0\,\text{s},\ t_1 = 5{\cdot}10^4\,\text{s},\ t_2 = 6{\cdot}10^4\,\text{s},\ t_3 = 11{\cdot}10^4\,\text{s},\ t_4 = 15{\cdot}10^4\,\text{s},\ t_4 = 20{\cdot}10^4\,\text{s}\,.$$

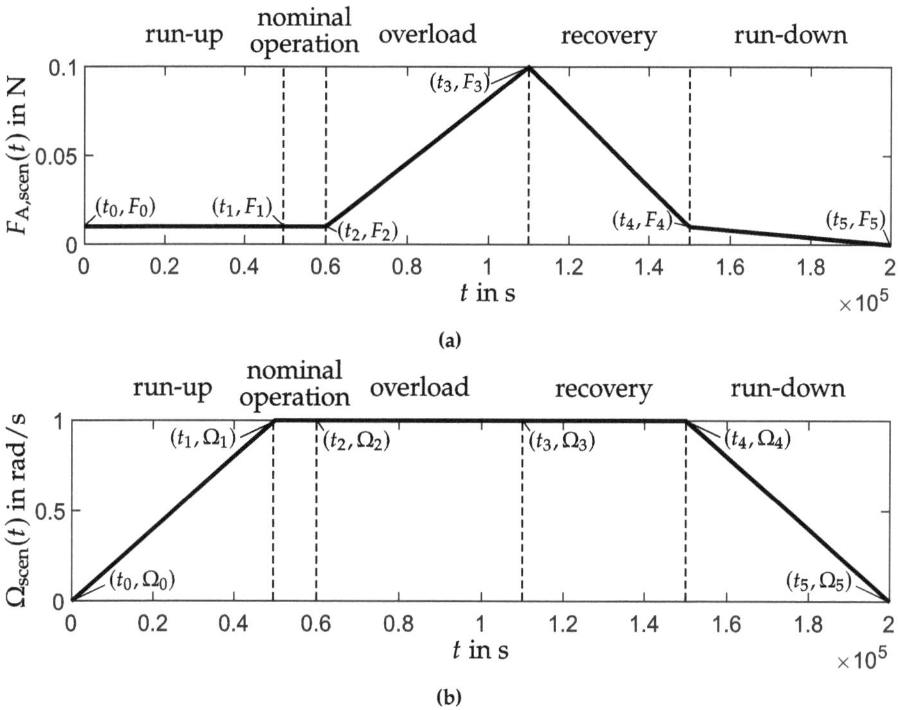

Figure 6.3: Components of the scenario excitation: (a) force amplitude $F_A(t)$ and (b) angular frequency $\Omega(t)$.

The initial conditions are given with the state vector $\mathbf{x}_0 = [x_{10}, x_{20}, \dot{x}_{10}, \dot{x}_{20}]^\top$. The initial conditions are chosen $\mathbf{x}_{0,\text{rect}} = [1, 1, 1, 1]^\top$ and $\mathbf{x}_{0,\text{sweep}} = \mathbf{x}_{0,\text{scen}} = [0, 0, 0, 0]^\top$ depending on the excitation. Furthermore, the constant parameters of the tuned wedge damper are

$$
\begin{aligned}
&m_1 = 1\,\text{kg},\ m_2 = 0.1\,\text{kg},\ c_1 = 1\,\text{N/m},\ c_2 = 0.1\,\text{N/m},\ c_3 = 0.01\,\text{N/m},\\
&\Delta_1 = 0.1\,\text{m},\ r = 0.01\,\text{m},\ \mu = 0.1\,.
\end{aligned}
$$

The angle and prestress displacement for the passive system are given by $\alpha_{\text{pa}} = 20^\circ$ and $\Delta\ell_{\text{pa}} = 0\,\text{m}$. The input parameters of the active system are limited to $\alpha \in [0, \alpha_{\max}]$ and $\Delta\ell \in [0, \Delta\ell_{\max}]$. The limits $\alpha_{\max}$ and $\Delta\ell_{\max}$ are specified for each control strategy later on and provide more realistic working conditions of the active tuned wedge damper.

6.2 Skyhook Control

As mentioned in section 2.3.1, the Skyhook Control emulates, when possible, a damper attached to an inertial frame, regardless of the actual damper position. Furthermore, this is a dissipation-oriented control strategy. To implement the Skyhook Control, it is first verified if the structure of dissipative forces is in accordance with the control strategy. Subsequently, the algorithm for the inputs is defined. Additionally, the input force is clipped to avoid sticking. Lastly, the section concludes with the simulations of the Skyhook Control behavior and its comparison to the passive system.

As noted in chapter 4, the dissipative terms of the tuned wedge damper are given by

$$F_{\mathrm{TWD,d}} = \frac{2\mu c_3(2y + \Delta\ell)(1 + y_x^2)}{1 - \mu y_x \operatorname{sign}(\dot{x}_2 - \dot{x}_1)} \operatorname{sign}(\dot{x}_2 - \dot{x}_1)\,. \tag{6.1}$$

The Skyhook Control assumes dissipative forces of the form $F_{\mathrm{Sky}} = |F_{\mathrm{TWD,d}}| \operatorname{sign}(\dot{x}_2 - \dot{x}_1)$, where the direction of the force is solely defined by the relative velocity. As is noted from Eq. (6.1), the dissipative force does not strictly comply with this form. This is due to the denominator, which can change signs depending on the relative velocity. However, for all practical purposes, this is not the case. Since the maximum input angle is limited to $\alpha_{\max} \leq 45°$, y_x has a maximum value of 1. Furthermore, since low friction coefficients are investigated ($\mu < 1$), it is not possible for the denominator to change signs. Taking into account the input limitations, the force in Eq. (6.1) complies with the necessary structure for the implementation of the control strategy.

The Skyhook Control defines a damper force that generates a braking effect on the main mass. This is only possible when the relative velocity between the masses and the absolute velocity of the main mass have different signs, i.e. $\dot{x}_1(\dot{x}_2 - \dot{x}_1) < 0$. When this is not the case, the damper force accelerates the main mass. To avoid exacerbating the oscillation of the main mass, the damper force is set to zero in these cases, i.e. when $\dot{x}_1(\dot{x}_2 - \dot{x}_1) > 0$. This results in the control structure

$$F_{\mathrm{TWD,d}} = \begin{cases} F_{\mathrm{TWD,d,max}}, & \dot{x}_1(\dot{x}_2 - \dot{x}_1) < 0 \\ 0, & \dot{x}_1(\dot{x}_2 - \dot{x}_1) > 0 \end{cases}\,. \tag{6.2}$$

The input parameters are calculated to ensure Eq. (6.2) and maximize the braking effect. For the wedge angle, this leads to a maximization of the input angle. The prestress displacement is maximized under the condition that sticking is avoided. If sticking occurs, no relative movement is observed, and, therefore, no energy is dissipated. Since this strategy is solely based on damping, sticking is counterproductive. The prestress displacement is thus calculated to comply with the slipping condition $|R| = \mu N > R_{\max}$, c.f. Eqs. (4.11) and (4.12). This yields for the wedge angle

$$\alpha_{\mathrm{Sky}} = \begin{cases} \alpha_{\max}, & \dot{x}_1(\dot{x}_2 - \dot{x}_1) < 0 \\ 0, & \dot{x}_1(\dot{x}_2 - \dot{x}_1) > 0 \end{cases} \tag{6.3}$$

and for the prestress displacement

$$\Delta\ell_{\mathrm{Sky}} = \begin{cases} \Delta\ell_{\max}(R, R_{\max}), & \dot{x}_1(\dot{x}_2 - \dot{x}_1) < 0 \\ 0, & \dot{x}_1(\dot{x}_2 - \dot{x}_1) > 0 \end{cases} . \tag{6.4}$$

To ensure smooth transitions between the extreme values, Eqs. (6.3) and (6.4) are regularized with the hyperbolic tangent function (tanh) and the regularization parameter $\varepsilon_{\mathrm{Sky}}$. This leads to

$$\alpha_{\mathrm{Sky}} = \frac{\alpha_{\max}}{2}\left(1 - \tanh\left(\frac{(\dot{x}_2 - \dot{x}_1)\dot{x}_1}{\varepsilon_{\mathrm{Sky}}}\right)\right), \tag{6.5}$$

$$\Delta\ell_{\mathrm{Sky}} = \frac{\Delta\ell_{\max}(R, R_{\max})}{2}\left(1 - \tanh\left(\frac{(\dot{x}_2 - \dot{x}_1)\dot{x}_1}{\varepsilon_{\mathrm{Sky}}}\right)\right). \tag{6.6}$$

Once the desired input parameters are known, the actual input parameters are changed accordingly. However, this change is not instantaneous, instead, the input variation is modeled dynamically with two first-order systems

$$\dot{\alpha} = -\lambda_{\mathrm{Sky}}(\alpha - \alpha_{\mathrm{Sky}}) \quad \text{and} \quad \dot{\Delta\ell} = -\lambda_{\mathrm{Sky}}(\Delta\ell - \Delta\ell_{\mathrm{Sky}}) \tag{6.7}$$

where λ_{Sky} is a measure of how fast the input parameters are changed. This input modeling approach has the advantage that it takes into account oscillations introduced by the change of the input parameters.

The parameters of the Skyhook Control Strategy are determined by trial and error and given by

$$\alpha_{\max,\mathrm{Sky}} = 20°,\ \Delta\ell_{\max,\mathrm{Sky}} = 240\,\mathrm{m},\ \lambda_{\mathrm{Sky}} = 0.64\,\mathrm{s}^{-1},\ \varepsilon_{\mathrm{Sky}} = 0.01\,\mathrm{m}^2/\mathrm{s}^2 .$$

Figure 6.4 compares the response to the impulse excitation of the Skyhook Control against its passive counterpart. The passive system introduces low dissipation which leads to a slow decay of the oscillations. In contrast, the Skyhook Control is able to introduce damping despite the low relative differences in the initial conditions. The control strategy dampens the free oscillations in the first 150 s. Furthermore, the rectangle impulse is countered effectively. The effect of the impulse on the main coordinate is minimal since in the pulse range it has a maximal value of 0.038 m. Once the excitation subsides, the amplitudes return to the vicinity of zero. At the end of the simulation, the amplitudes of the Skyhook Control are zero, whereas the amplitudes of the passive system are roughly 0.500 m.

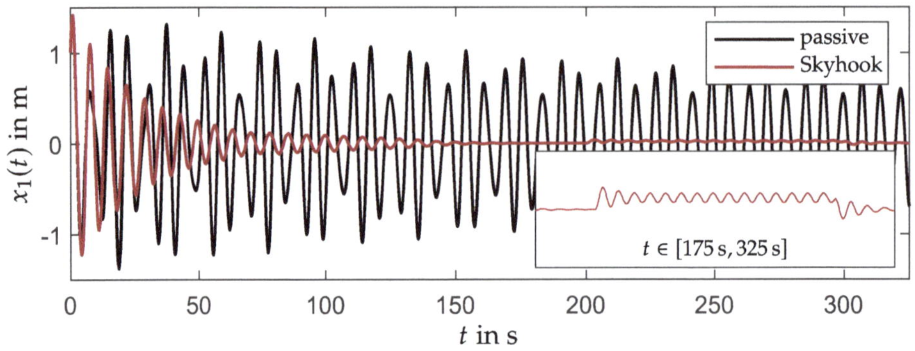

Figure 6.4: Response of the Skyhook Control to the rectangular pulse.

The sweep response of the Skyhook Control is observed in Fig. 6.5. The amplitudes of both systems are approximately equal for frequencies outside the resonance and absorption regimes. Furthermore, the maximum amplitudes of the active damper are substantially lower than its passive counterpart. Compared to the maximum value of 2.060 m of the passive damper, the active damper has a maximum amplitude of 0.341 m, which represents a reduction of 83.57 %. In the vicinity of the absorption frequency, the disadvantages of the Skyhook Control are noted. Since this strategy is dissipation-focused, the amplitudes are higher than its passive counterpart. At the absorption, the active damper amplitude has a value of 0.046 m. After the absorption regime the amplitudes of the active damper rise again, however, they remain under 0.101 m. This rise is due to the resonance regime of the second eigenfrequency.

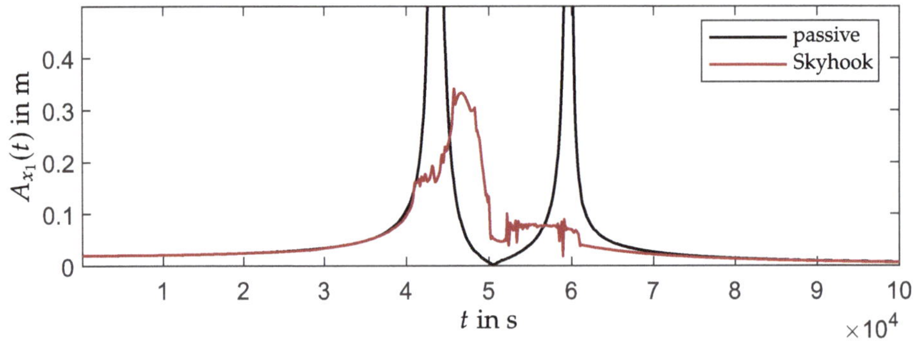

Figure 6.5: Response of the Skyhook Control to the sweep excitation.

The response to the application scenario is presented in Fig. 6.6. During the run-up phase, the active system shows a lower maximum amplitude than its passive counterpart. The disadvantages of the active system show starting from the nominal operation phase. Due to the focus on damping, the oscillation amplitudes are not minimal during this phase. Instead, they have a value of 0.049 m. Additionally, the Skyhook Control shows different amplitude ranges during the overload phase. These are caused by the changes in the maximum input values that the control allows. Due to the increased amplitude of the excitation, the maximum stiction force increases and with it the maximum value of the prestress displacement. The added possibility to introduce more damping leads to the different ranges in the response and the transition between them. In the overload phase, a maximum amplitude of 0.580 m is observed, which is by a factor 12.95 higher than the passive system. The passive system takes advantage of absorption and therefore has lower amplitudes. Nevertheless, over the whole scenario, the maximum amplitudes of the Skyhook Control are 45.48 % lower than those of the passive system. During the recovery phase the amplitudes subside to the nominal operation levels. During the run-down the system passes once again through the resonance regime and the amplitudes rise and have a value of 0.156 m. Compared to the passive system during run-down this represents a 82.65 % amplitude reduction.

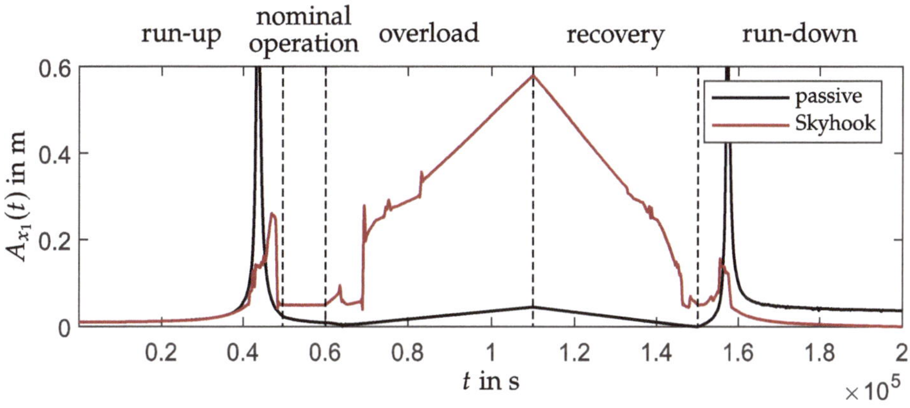

Figure 6.6: Response of the Skyhook Control to the application scenario.

6.3 Adaptive Multiple Model Control

The adaptive multiple model control (MMC) is based on the adaptive control with a reference model. This strategy combines different vibration reduction models to achieve an optimal response. Accordingly, this control strategy does not focus solely on damping, but instead also uses absorption and changes in the system's eigenfrequencies via sticking. First, the general control structure is presented. Second, the individual actuation models are described. Third, the decision criterion of the supervising controller that chooses the applied actuation model is formulated. Lastly, the simulations of the multiple model control are presented and compared to the passive system.

The proposed control is based on the investigations of Morse [128] and Narendra and Balakrishnan [129]. In their work, they extended the classical adaptive control with a reference model by using several reference models with their corresponding controllers. All models are simulated in parallel and identify the plant simultaneously. The method described in this work has in common with [129] that several models are simulated at the same time. The structure of the decision criterion is also partially taken from this work. The difference in approach is that the adaptive multiple model control does not perform an identification. Instead, the control tries, similar to [128], to determine which input variables of the models cause the greatest vibration reduction. For this purpose, three models are simulated in parallel. These models represent the mechanisms for vibration reduction: dissipation, absorption, and shifting of structural resonances via sticking. Instead of reference models, this control uses actuation models.

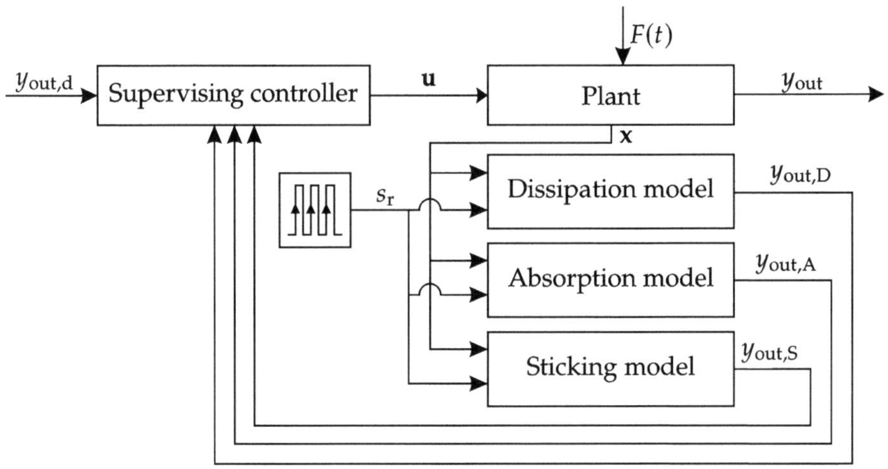

Figure 6.7: Control structure of the adaptive multiple model control.

The control structure is presented in Fig. 6.7. The states of the system $\mathbf{x} = [x_1, x_2, \dot{x}_1, \dot{x}_2]^\top$ are forwarded to each actuation model. Additionally, they receive the reset signal s_r. Depending on the period of the reset signal T_r, the states of the models are reset to the states of the plant with every rising edge. Resetting the states of the actuation models is necessary so that they can provide a relevant prediction about the system behavior. Otherwise, the actuation models that are not in use will drift off. Furthermore, the actuation models simulate the effect of their parameter choice, taking into account any stick-slip transitions, but without knowledge of the perturbation $F(t)$. The outputs of the models $y_{out,D}$, $y_{out,A}$, and $y_{out,S}$ are fed back to the supervisory controller, which then decides which vibration reduction approach is accepted. In Fig. 6.7, the classic control and adaptation loops cannot be separated from each other. However, the adaptation law is noted in the supervising controller. The controller changes its approach depending on the simulated performance of the actuation models. In this structure, the control and adaptation loops merge into one branched loop. Although one actuation model is always active, the system has short intervals where it changes from a closed-loop to an open-loop control. This is the case when it changes the actuation model. At this moment, the desired system input and the actual system input are not equal and therefore differences arise between the actuation model and the plant. However, the strategy returns to a closed-loop control once these differences subside or the states are reset.

The dissipation model adjusts the parameters according to the Skyhook Control in section 6.2. If the relative velocity between the masses and the absolute velocity of the main mass have different signs, damping is maximized. The angle and the prestress displacement take their maximum values. Sticking, however, is avoided. If the signs of the considered velocities are the same, the parameters are set to zero. This avoids exacerbating the vibrations of the main mass.

The absorption model adjusts the parameters so that absorption is achieved. Both the angle and the prestress are set to zero. As such the dissipation brought on by the damper is eliminated. The system behaves like a vibration absorber with the secondary mass m_2 and the secondary stiffness c_2. This model is particularly relevant near the absorption frequency $\omega_{\text{abs}} = \sqrt{c_2/m_2}$.

The sticking model simulates the system as if it were forced to stick. To this end, the wedge angle $\alpha = 0$ and the prestress displacement $\Delta\ell = \Delta\ell_{\text{max}}$ are set. Sticking changes the structure of the system. The two eigenfrequencies of the slipping system vanish and only the resonance frequency of the sticking system remains. This change is used to avoid the structural resonance frequencies.

Based on the parameter suggestions of the actuation models, the supervisory controller decides which of these candidates is best for vibration reduction. For this purpose, each model is assigned a performance index. The performance index J_i is calculated using the performance function $p_i(\mathbf{x})$ with $i = \{\text{D}, \text{A}, \text{S}\}$. These are given by

$$J_i = \gamma p_i(t, \mathbf{x}) + \beta \int_{nT_r}^{t} \mathrm{e}^{-\lambda_f(t-\tau)} p_i(\tau, \mathbf{x}) \mathrm{d}\tau \quad \text{and} \quad p_i(\mathbf{x}) = \left| y_{\text{out,d}} - y_{\text{out},i} \right| . \tag{6.8}$$

In Eq. (6.8), γ and β are the weight factors of the terms. The first term represents the current value of the performance function, while the second term corresponds to the value of the performance function in the past. To avoid considering irrelevant dynamics of the past, the integral only takes into account the time interval $[nT_r, t]$ transgressed since the last state reset with $n = \lfloor t/T_r \rfloor$. Additionally, an exponential forgetting with the forgetting factor λ_f determines how the individual values of the past are weighted in the integral. The underlying performance function represents the absolute value of the difference between the desired output $y_{out,d}$ and the output of the respective actuation model $y_{out,i}$. Since vibration reduction of the main mass is of interest, the position and velocity of the main mass are implemented in the performance functions with

$$y_{out,i} = \sqrt{x_1^2 + \dot{x}_1^2} \tag{6.9}$$

and $y_{out,d} = 0$. The actuation model with the minimum performance index is chosen and its value for the wedge angle and the prestress displacement are set as the desired input parameters α_{MMC} and $\Delta\ell_{MMC}$. As with the Skyhook Control, the effect of changing the input variables is represented by the differential equations

$$\dot{\alpha} = -\lambda_{MMC}(\alpha - \alpha_{MMC}) \quad \text{and} \quad \dot{\Delta\ell} = -\lambda_{MMC}(\Delta\ell - \Delta\ell_{MMC}) \,. \tag{6.10}$$

Analogously, the parameter λ_{MMC} in Eq. (6.10) is a measure of the speed at which the input variables are changed. The advantage of this strategy is that it not only takes into account the momentary influence of the damper but also its past performance. Additionally, it is not limited to a single vibration reduction mechanism. Depending on the parameter choice, the time frame of the control strategy is defined. Large rates of change λ_{MMC} and small reset periods T_r lead to a strategy that acts on a small time frame. It is constantly updating and acts within the lapse of a single oscillation. Low rates of change and large reset periods result in a strategy that acts on a larger time frame.

The multiple model control parameters are derived by trial and error. The control parameters are chosen to minimize oscillations while achieving stationary solutions. For this strategy, parameters are given by

$$\alpha_{max,MMC} = 20^\circ, \; \Delta\ell_{max,MMC} = 10\,\text{m}, \; \lambda_{MMC} = 5\text{e}{-4}\,\text{s}^{-1}, \; \gamma = 0, \; \beta = 1,$$
$$\varepsilon_{Sky} = 0.010\,\text{m}^2/\text{s}^2, \; \lambda_f = 0.126\,\text{s}^{-1}, \; T_r = 12.566\,\text{s} \,.$$

The results for the rectangular pulse excitation are presented in Fig. 6.8. In the first 200 s, the multiple model control reduces the free oscillations of the system, however, small oscillations with an amplitude of 0.122 m remain. During the pulse, the vibrations of the main mass oscillate around 0.018 m with an amplitude of 0.139 m. After the pulse excitation, small oscillations remain and have an amplitude of 0.123 m. Compared to the passive systems this results in a 79.53 % reduction in the vibration amplitude. In contrast to the Skyhook Control, the system oscillations are not brought to a halt. Therefore, a parameter or strategy change is sensible for small oscillations.

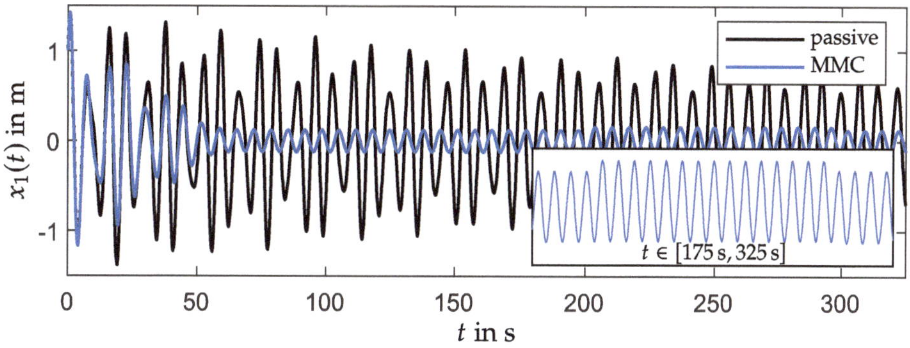

Figure 6.8: Response of the adaptive multiple model control to the rectangular pulse.

Figure 6.9 presents the response of the multiple model control to a sweep excitation. The maximum amplitudes of the sweep response have a value of 0.247 m. This equates to an 88.08 % reduction compared to the passive system. Disadvantages of this control strategy appear in the absorption range where the optimal solution via absorption is not recognized. Although complete absorption is not observed, the amplitudes remain low. At the absorption frequency, an amplitude of 0.052 m is observed.

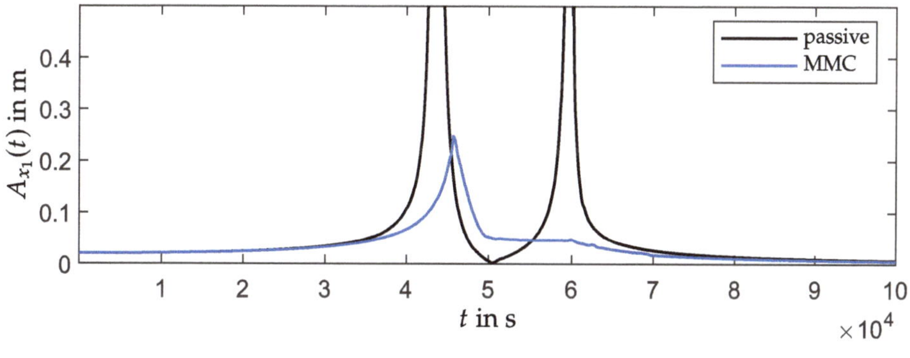

Figure 6.9: Response of the adaptive multiple model control to the sweep excitation.

The response of the multiple model control to the application scenario is presented in Fig. 6.10. In the run-up phase, the control strategy has a maximum amplitude of 0.248 m, which corresponds to a 76.72 % reduction compared to the passive system. During the nominal operation, the passive system responds with lower amplitudes than the active control strategy. As with the sweep excitation, the control strategy does not recognize the amplitude reduction potential via absorption. However, the amplitudes are low and have a value of 0.037 m. Throughout the overload phase, the amplitudes rise and attain a maximum value of 0.156 m. Although the excitation plays an increasing role in the system's dynamics and the actuation models do not simulate it, a substantial increase in the amplitude of the main mass is not observed. Even though the excitation increases tenfold, the maximal amplitudes of the overload phase are only 4.20 times lager than the amplitude during the nominal operation phase. Compared to the passive system, the maximal overload amplitude is higher by a factor of 3.49. The oscillation amplitudes fall to 0.034 m at the end of the recovery phase. During the passage through resonance of the run-down phase the resonance peak is lower than during the run-up phase and has a value of 0.101 m, which is 88.75 % lower than its passive counterpart.

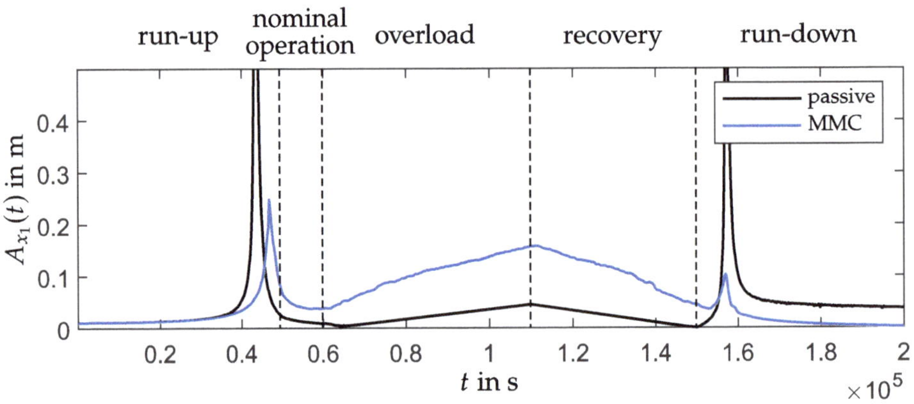

Figure 6.10: Response of the adaptive multiple model control to the application scenario.

6.4 Slow Frequency-Based Control

The slow frequency-based control (SFC) is similar to the gain scheduling control strategies. It uses prior knowledge of the systems and defines the system parameters accordingly. In this case, the prior knowledge used is based on the frequency response of the tuned wedge damper. To describe the slow frequency-based control, first, the ideal change of the control strategy is described. Second, the control structure, as well as the necessary frequency identification, is presented. Third, the rule for the parameter choice is formulated. Lastly, the simulations of the active control strategy are presented.

This control method uses knowledge of the frequency response to determine the system parameters. The two frequency responses and an ideal case, are plotted in Fig. 6.11. The figure shows the responses of the sticking ($\Delta\ell \to \infty$) and slipping ($\Delta\ell = 0$) systems. The sticking response is modeled with Eq. (4.10), and the slipping response corresponds to

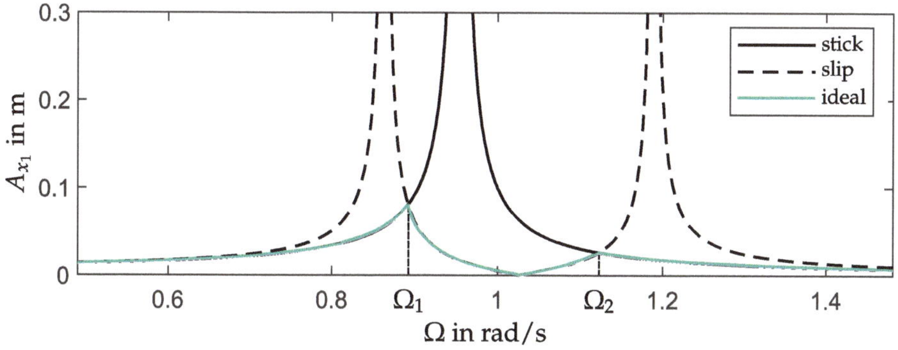

Figure 6.11: The ideal change between the sticking and slipping frequency response functions.

an undamped tuned mass damper with the secondary stiffness $c_{\text{eff,res}} = c_2 + 4c_3 \tan^2 \alpha$. The ideal response is built by selecting the branch with the lowest amplitude. The ideal response has the advantages of low vibration amplitudes without having to forego an absorption frequency. The slow frequency-based control method aims to reproduce this ideal curve. The underlying basic principle of the system is the change in the system's structure caused by the transitions between sticking and slipping. As the system transitions, the eigenfrequencies change from two slipping eigenfrequencies to a single sticking frequency or vice versa. The targeted change aims to avoid resonances and achieve absorption.

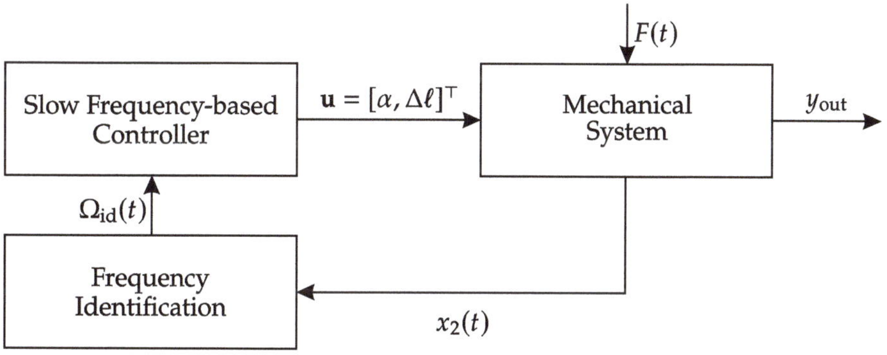

Figure 6.12: The control structure of the slow frequency-based control.

The overall system structure is shown in Fig. 6.12 and consists of three blocks: the mechanical system block, the frequency identification block, and the controller block. To influence the mechanical system, the two input parameters α and $\Delta\ell$ are varied. These two parameters serve as inputs to the mechanical system. This block contains and simulates the equations of motion of the tuned wedge damper with the corresponding parameters. The resulting movement of the auxiliary mass $x_2(t)$ is recorded and passed to the frequency identification block. This variable is chosen since it does not tend to zero in the absorption regime. This simplifies the identification. The identification block detects the main frequency of the system using a FOURIER Transformation. The identified frequency Ω_{id} is then passed on to the controller, which sets the parameters α and $\Delta\ell$. The idea of this controller is based on the adaptive gain scheduling considered in section 2.3. It is similar in the sense that an additional quantity, the identified frequency, is used, to gain additional knowledge of the system. However, it is noted that the structure in Fig. 6.12 is that of the open-loop control, since the output is not fed back to the system. Furthermore, as is stated later on, the strategy does not change the parameters of a controller, but instead directly defines the desired input parameters.

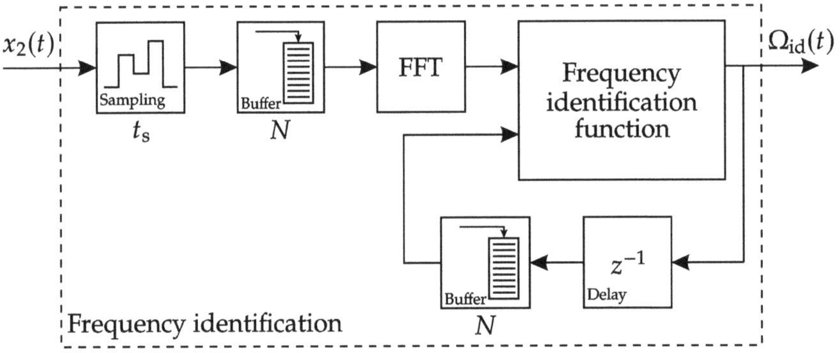

Figure 6.13: Structure of the frequency identification block

A detailed view of the identification block is shown in Fig. 6.13. The measurement of $x_2(t)$ is sampled with the sampling time t_s and the last N points are collected in a time series. A Fast FOURIER Transformation (FFT) of the time series is calculated. Subsequently, the oscillation frequency is determined by the frequency identification function. This function establishes the frequencies of the FFT peaks as possible frequency candidates. The candidates are compared to the moving average of the identified frequency so far. The oscillation frequency is identified as the candidate with the smallest distance to the moving average. This frequency is then forwarded to the controller.

The controller chooses the desired input values α_{SFC} and $\Delta\ell_{SFC}$ according to

$$\alpha_{SFC} = \begin{cases} \alpha_{max,SFC}, & \Omega_1 \leq \Omega_{id} \leq \Omega_2 \\ 0, & \text{else} \end{cases} , \tag{6.11}$$

$$\Delta\ell_{SFC} = \begin{cases} 0, & \Omega_1 \leq \Omega_{id} \leq \Omega_2 \\ \Delta\ell_{max,SFC}, & \text{else} \end{cases} . \tag{6.12}$$

In Eqs. (6.11) and (6.12), Ω_1 and Ω_2 represent the crossings frequencies between the linear sticking and slipping systems, see Fig. 6.11. Outside the Ω_1-Ω_2-range, the lowest amplitudes are achieved by the sticking system. Thus, the system parameters are chosen so that the tuned wedge damper sticks. The outer segments are flat and the prestress displacement is maximized. In the Ω_1-Ω_2-range, absorption is the chosen vibration reduction mechanism. To this end, the prestress is set to zero to minimize damping. Additionally, the wedge angle α is maximized to insert damping outside the $2\Delta_1$-range. This parameter choice funnels large oscillations into the dissipation-free range. The control strategy highly depends on the identified frequency since the parameters are solely determined by it. This is problematic in the presence of noise in the measured signal. This is manageable by applying a low pass filter to the measured signal provided the magnitude of the noise is small compared to $x_2(t)$. The frequencies Ω_1 and Ω_2 are calculated based on the crossings of the linear systems in Fig. 6.11. These are independent of the excitation amplitude and are given by

$$\Omega_{1,2} = \omega_{01}\sqrt{\frac{\gamma + (1+\gamma)p^2 \mp \sqrt{(1+\gamma)^2p^4 - 2\gamma p^2 + \gamma^2}}{\gamma(2+\gamma)}} \tag{6.13}$$
$$\text{with } \gamma = \frac{m_2}{m_1},\ p^2 = \frac{c_{2,\text{eff}}}{c_1},\ \omega_{01}^2 = \frac{c_1}{m_1},\ c_{2,\text{eff}} = c_2 + 4c_3\tan^2\alpha_{\text{max,SFC}}\,.$$

As with the aforementioned strategies a dynamic input model is considered. With the rate of change for the inputs λ_{SFC}, the differential equations for the inputs are given by

$$\dot{\alpha} = -\lambda_{\text{SFC}}(\alpha - \alpha_{\text{SFC}}) \quad \text{and} \quad \dot{\Delta\ell} = -\lambda_{\text{SFC}}(\Delta\ell - \Delta\ell_{\text{SFC}})\,. \tag{6.14}$$

The slow frequency-based control does not aim to counter vibration in the lapse of one oscillation. Instead, it focuses on the attenuation of vibrations over a large time frame. It is therefore important that the changes of the parameter don't introduce high frequency oscillations in the system since the control strategy cannot promptly react to these. To this end, the rate of change λ_{SFC}, the maximum wedge angle $\alpha_{\text{max,SFC}}$, and the maximum prestress displacement $\Delta\ell_{\text{max,SFC}}$ are chosen as small. However, not unreasonably small that vibration reduction is not realized. The first two parameters are found by trial and error. The maximum prestress value is approximated with the transition estimate in Eq. (4.18). From the parameters of the passive system and the amplitude of the excitation, the maximum value results in

$$\Delta\ell_{\text{max,SFC}} = \frac{A_L(\Omega_1)\Omega_1^2 m_2}{2c_3\mu} \quad \text{with} \quad A_L(\Omega_1) = \frac{F_0}{|c_1 - (m_1+m_2)\Omega_1^2|}\,. \tag{6.15}$$

Although the parameter F_0 is found in Eq. (6.15), the exact value is not required. Equation (6.15) is meant as an estimate for the maximum prestress value, accordingly an estimate of the magnitude order of F_0 is sufficient. The expression above yields the minimal prestress displacement to ensure that the system sticks until Ω_1.

For the identification, the sampling parameter t_s and the series length N have to be counterbalanced. These parameters determine the maximal identified frequency $\Omega_{id,max}$ as well as the minimal change of the identified frequency $\Delta\Omega_{id}$. The latter affects the resolution of the FFT. With the NYQUIST-SHANNON criterion, these values are given by

$$\Omega_{id,max} = \frac{\pi}{t_s} \quad \text{and} \quad \Delta\Omega_{id} = \frac{2\pi}{N t_s}. \tag{6.16}$$

Low sampling times t_s lead to a larger frequency range, however, they also cause low resolution in the FFT, i.e. large $\Delta\Omega_{id}$ values. This is especially impairing in the vicinity of the transition frequencies since the targeted stick-slip transition is not executed accurately. A high time series length N helps improve the resolution of the FFT. However, it increases the initial buffer time $N t_s$ in which the oscillation frequency is not correctly identified. These inaccuracies appear because at first the buffer in the identification block is mostly filled with zeros. Low N values, therefore, shorten this time and improve the speed with which correct frequencies are identified. A counterbalancing of all these properties is needed. However, since this control strategy orients itself to larger time frames, the covering of the required frequency range and the resolution of the FFT have priority over the initial buffer time.

The parameters for the slow frequency-based control are chosen as

$$\alpha_{max,SFC} = 20°,\ \Delta\ell_{max,SFC} = 3.219\,\text{m},\ \lambda_{SFC} = 0.001\,\text{s}^{-1},\ F_0 = 0.01\,\text{N},$$
$$N = 256,\ t_s = 1\,\text{s},\ \Omega_1 = 0.892\,\text{rad/s},\ \Omega_2 = 1.122\,\text{rad/s}.$$

The behavior in response to the rectangular impulse is presented in Fig. 6.14. Although the control strategy is not designed for the transient response, it effectively reduces the oscillation amplitudes. The amplitudes nearly vanish in the first 200 s of the simulation. During the rectangular pulse, the amplitudes rise to 0.080 m. At the end of the simulation, an amplitude reduction of 91.83 % is achieved. However, small oscillations remain in the active system and have an amplitude of 0.049 m.

Figure 6.15 depicts the sweep response of the slow frequency-based control. Although the ideal change is not precisely realized, the control strategy has an advantageous response. It follows the sticking system on the outer frequency range and in the vicinity

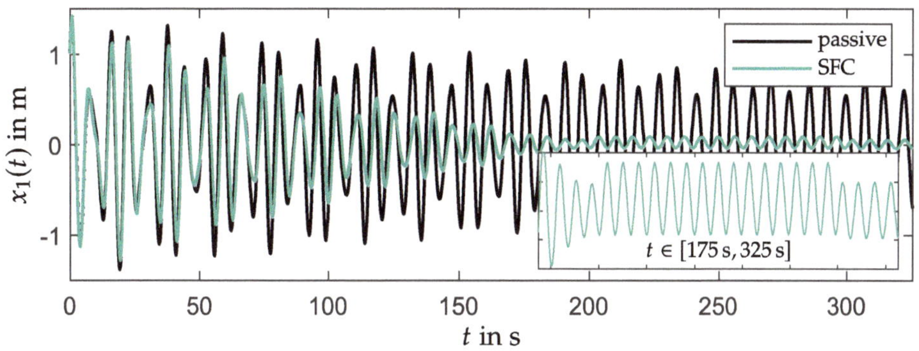

Figure 6.14: Response of the slow frequency-based control to the rectangular pulse.

of the absorption it follows the passive system. Due to the changes in the wedge angle during the transition between the sticking and slipping systems, oscillations in the maximum amplitude are noted. Furthermore, a maximum amplitude of 0.180 m is noted. This results in a 91.11 % amplitude reduction compared to the passive system. With the targeted stick-slip transition, the resonance frequencies are avoided and the advantages of sticking and slipping are combined.

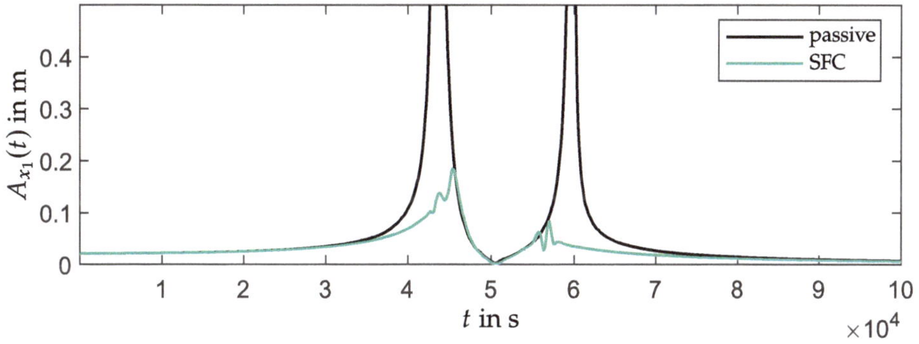

Figure 6.15: Response of the slow frequency-based control to a sweep excitation.

The response of the slow frequency-based control to the application scenario is presented in Fig. 6.16. The strategy results in a significant improvement on the passive system. During the run-up phase, the control strategy shows a maximum amplitude of 0.154 m, which is 85.56 % lower than the passive system. Since the maximum amplitude of the passive system is larger, it is not able to realize full absorption during the nominal operation range. In contrast, the amplitudes of the slow frequency-based control are in the vicinity of zero. Both systems show the same response during the overload and recovery phase and have a maximum amplitude of 0.045 m. During the run-down phase, the oscillation amplitudes rise again. The maximum amplitude in this range is higher than during the run-up phase and has a maximum value of 0.247 m. This amplitude equates to a 75.55 % amplitude reduction compared to its passive counterpart.

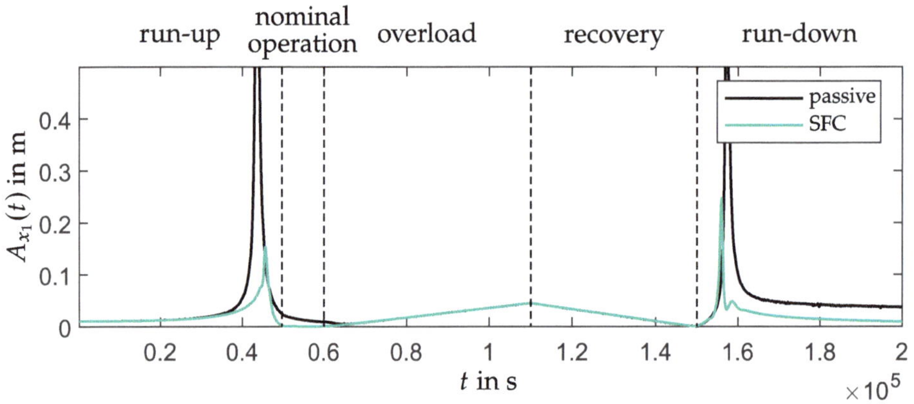

Figure 6.16: Response of the slow frequency-based control to the application scenario.

6.5 Control Strategy Comparison

After investigating the control strategies separately, the focus is set on their comparison. The comparison is carried out according to the aforementioned excitations. First, the rectangular pulse is considered. Second, the sweep excitation is taken into account. Finally, the strategies are compared in the application scenario.

Figure 6.17 portrays the comparison of the control strategies in response to the rectangular pulse. The Skyhook Control shows the best results since its approach attenuates vibrations faster. Additionally, it brings the system to a halt. In the beginning,

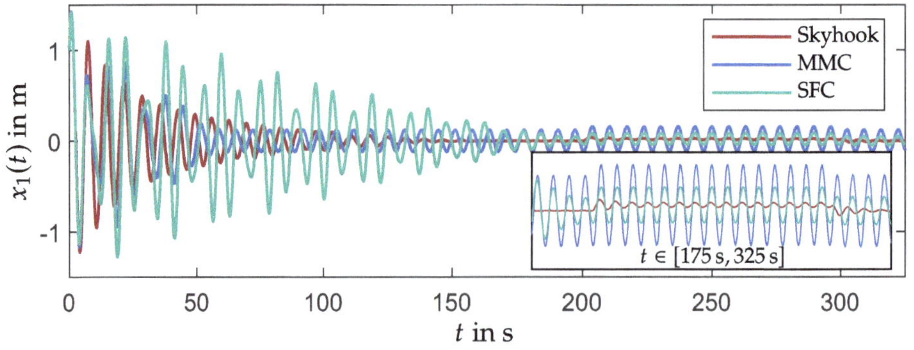

Figure 6.17: Comparison of the control strategies for the rectangular pulse.

the adaptive multiple model control reduces the oscillations as well as the Skyhook Control. However, at the end small oscillations remain. Nevertheless, it has the advantage that it does not solely rely on damping. Therefore, a longer life span is expected with the multiple model control. The slow frequency-based control takes longer to mitigate the vibrations. However, large oscillations are reduced even though this strategy was not designed for transient behavior. The amplitude of the Skyhook Control in reaction to the rectangular pulse is smaller than the other control strategies by a factor of 2.16 and 4.16 for the slow frequency-based control and the multiple model control, respectively. In the end, the multiple model control and the slow frequency-based control do not bring the system to a halt and small oscillations remain.

The comparison of the active control strategies' response to the sweep excitation is presented in Fig. 6.18. The slow frequency-based control yields the best response. It has a maximal amplitude of 0.180 m which is 46.14 % lower than the Skyhook Control and 25.65 % lower than the multiple model control. Furthermore, in the absorption range, the slow frequency-based control achieves the lowest amplitudes. It uses vibration absorption as its primary vibration reduction mechanism and only dissipates when the oscillations are partly outside the $2\Delta_1$-range. Outside the resonance and absorption regimes, the control strategies yield approximately the same results.

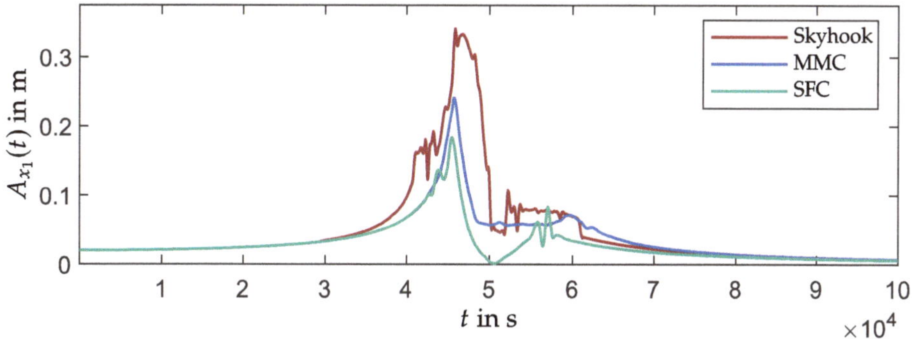

Figure 6.18: Comparison of the control strategies for the sweep excitation.

Figure 6.19 shows the comparison of the active control strategies in the application scenario. Due to the higher velocity, with which the structural resonances are passed through, the maximum amplitudes during the run-up phase are smaller than in Fig. 6.18. However, the qualitative behavior of the control strategies remains the same. The slow frequency-based control still shows the lowest amplitudes. This establishes the advantages of absorption and shifting structural resonances via sticking as effective vibration reduction mechanisms. The second-best results are shown by the multiple model control and the highest amplitudes are shown by the Skyhook Control. This is noted from the run-up to the recovery phase and confirms that dissipation is not always the best vibration reduction strategy. During the overload phase, the slow frequency-based control achieves the lowest maximal amplitudes of 0.045 m. These are 92.28 % smaller than the Skyhook Control and 71.32 % smaller than the multiple model control. The slow frequency-based control only shows a suboptimal response during the run-down phase. During this phase, the oscillations of the slow frequency-based control are 1.58 times higher than the Skyhook Control and 2.44 times higher than the multiple model control. Considering that the slow frequency-based control shows overall the lowest amplitudes, its behavior represents the optimal response.

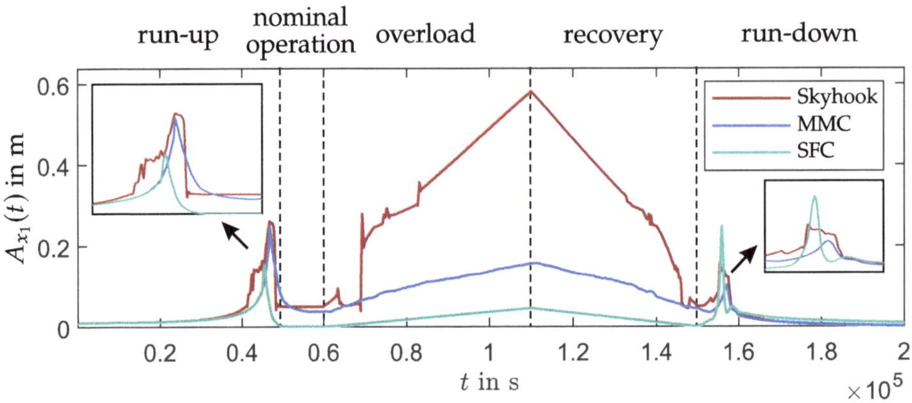

Figure 6.19: Comparison of the control strategies for the application scenario.

6.6 Energy Investigations

The active control strategies introduce dissipation in a more targeted manner compared to the passive system. This targeted dissipation yields lower amplitudes, however, too much dissipation results in a shorter service life. Therefore, a counterbalancing between amplitude reduction and life span is required. Since the control strategies suggested in this work do not solely rely on dissipation, they are able to reduce vibrations with a lower dissipated energy. Thus, achieving a longer damper service life and energy savings with low vibrations amplitudes. To assess these qualities, the dissipated energy and the input energy of the control strategies are investigated. First, the formulas for the dissipated energy is derived. Second, the expression for the input energy is formulated. Afterward, the formulas are used to evaluate the strategies in the three excitation scenarios.

The calculation of the dissipated energy $E_{\text{TWD,d}}(t)$ is derived from the dissipated power $P_{\text{TWD,d}}(t)$. The latter is calculated from the product of the dissipative force $F_{\text{TWD,d}}$ and the relative velocity $\dot{x}_{\text{rel}} = \dot{x}_2 - \dot{x}_1$. The dissipated energy is associated with the wear of materials. Increasing levels of dissipated energy lead to higher wear volume [161]. This is especially relevant for friction dampers since surfaces are often rubbing against each other. In the optimal case, dampers are able to reduce vibration with low $E_{\text{TWD,d}}(t)$-values. Lower values yield a longer damper life span. Considering Eq. (6.1) and the time-dependent values $\alpha(t)$ and $\Delta\ell(t)$, the dissipative power results in

$$P_{\text{TWD,d}}(t) = -F_{\text{TWD,d}}(t)(\dot{x}_2 - \dot{x}_1) = -\frac{2\mu c_3\big(2y(t) + \Delta\ell(t)\big)\big(1 + y_x^2(t)\big)}{1 - \mu y_x(t)\,\text{sign}(\dot{x}_2 - \dot{x}_1)}|\dot{x}_2 - \dot{x}_1|\,. \quad (6.17)$$

The dissipated power is always negative since mechanical energy is taken from the system. This is confirmed by the terms above since the fraction in Eq. (6.17) is always positive, as noted in section 6.2. From the expression above, the dissipated energy is calculated by integrating the dissipated power above over time. This results in

$$E_{\text{TWD,d}}(t) = \int_0^t P_{\text{TWD,d}}(\tau)\,\mathrm{d}\tau\,. \quad (6.18)$$

Since concrete designs for the actuators are not considered, a conservative estimate for the minimal energy consumption is approximated. To calculate the approximation, the minimal mechanical power required by the wedge angle and the mechanical power required by the prestress displacement are approximated. The input power of the wedge angle is estimated by considering the minimal torque M_α needed to rotate the outer flanks. This is determined by the minimal force $\tilde{N}$ to overcome the normal force N and the lever arm ℓ_α, see Fig. 6.20a. The pivot point is assumed at the transition from the horizontal segments to the circular segments. The product of the moment

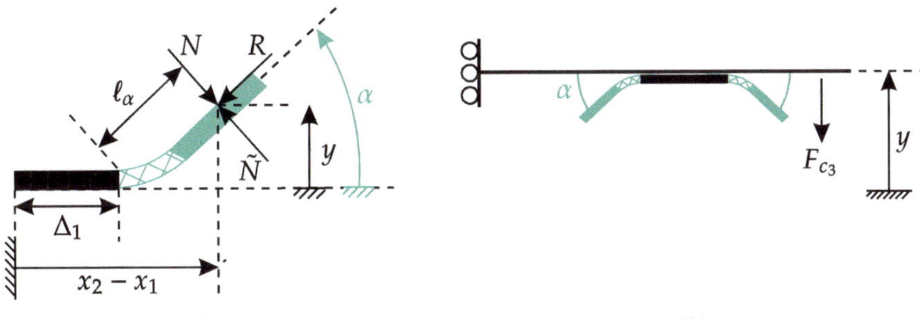

Figure 6.20: Free body diagrams for the calculation of the necessary power and energy of (a) the wedge angel α and (b) the prestress displacement $\Delta\ell$.

$M_\alpha = \tilde{N}\ell_\alpha = N\ell_\alpha$ and the angular velocity of the outer segments $\dot{\alpha}$ yields the power associated with the angle α. Since the prestress displacement is associated with a portion of the spring force F_{c3}, the mechanical power associated with this input is directly calculated. The input power of the prestress displacement is determined by considering the force $F_{\Delta\ell} = c_3\Delta\ell$ and the velocity with which the surfaces are pressed apart $\dot{y}$, see Fig. 6.20b. Since the circular transition segments are small, the lever arm is approximated with

$$\ell_\alpha = \frac{1 + \text{sign}(|x_2 - x_1| - \Delta_1)}{2}\sqrt{(|x_2 - x_1| - \Delta_1) + y^2}\,.$$

The dissipated power is thus given by

$$P_{\text{inputs}} = P_\alpha + P_{\Delta\ell} = 2\max(M_\alpha\dot{\alpha}, 0) + 2\max(\text{-}c_3\Delta\ell\dot{y}, 0)\,. \tag{6.19}$$

Analogous to the dissipated energy, the input energy results from

$$E_{\text{inputs}}(t) = \int_0^t P_{\text{inputs}}(\tau)\,\mathrm{d}\tau\,. \tag{6.20}$$

Equation (6.20) has three implicit assumptions. First, since only the mechanical power is used as the basis for the calculation, energy is only introduced with a displacement of the contact surfaces or the rotation of the outer surfaces. Energy consumption for the holding of a position is not considered. This is accurate for a self-locking design. Furthermore, an increase in the input energy due to prestress changes, while $\dot{y} = 0$, is not considered since, again, only the mechanical power is taken into account. Second, the use of the max-function considers only positive power values. Therefore, energy recovery is not considered, yielding a more conservative estimate. Third, the power associated with the angle considers only the two flanks in contact and only the minimal resistance force. The additional force required to move the flanks with a defined progression and the energy needed to move the flanks, not in contact, are not considered. Nevertheless, Eq. (6.20) represents a lower limit for the input energy of the control strategies. Furthermore, it allows the evaluation of the input energy of the strategies in the absence of a specific actuator model.

The equations derived above are used to evaluate the control strategies with respect to dissipated energy and input energy. First, the evaluations regarding the dissipated energy are made for the rectangular pulse, the sweep excitation, and the application scenario. Equation (6.18) is also applied to the passive system and compared to the semi-active control strategies. Subsequently, the evaluations for the input energy are presented for the control strategies and the aforementioned excitation.

Figure 6.21 shows the dissipated energy of the passive damper and the control strategies for the rectangular pulse. The Skyhook Control Strategy dissipates 1.421 J and thus dissipates the most energy. Furthermore, the dissipation is introduced in the first 106 s in a targeted manner. The second fastest control strategy is the multiple model control which dissipates 1.043 J in 60 s. The slow frequency-based control dissipates slightly more energy, 1.046 J, than the multiple model control. However, it requires 191 s, resulting in a slower strategy. The passive system does not reach a saturation level and at the end of the simulations it has the lowest dissipated energy, namely 0.731 J. Nevertheless, the main goal, vibration reduction, is not achieved since its vibration amplitudes are much higher than the active variants, c.f. Figs. 6.4, 6.8 and 6.14. The curves show that the control strategies are able to introduce damping in a more targeted manner, which results in lower amplitudes in a shorter time. Taking the passive system at the end of the simulation as a reference point, the dissipated energy of the multiple model control, the slow frequency-based control, and the Skyhook Control are respectively 42.78 %, 43.06 %, and 94.42 % higher than the passive system.

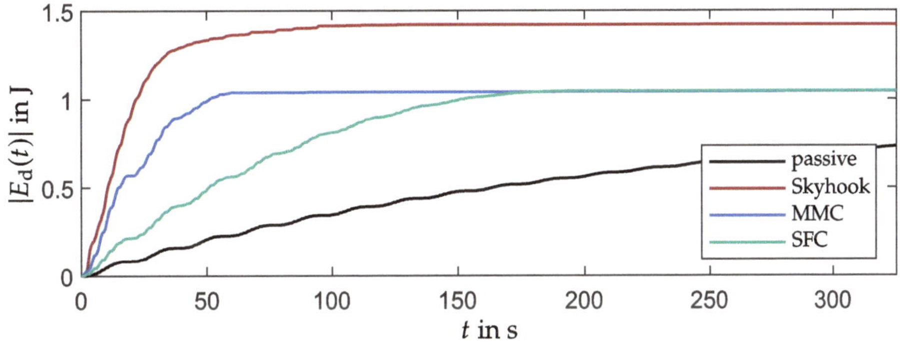

Figure 6.21: Comparison of the dissipated energy for the rectangular pulse.

The dissipated energy for the sweep excitation is shown in Fig. 6.22. Once again the Skyhook Control shows the highest dissipated energy. Over the whole sweep range, the strategy dissipates 21.365 J. The highest increase in dissipated energy is observed between the two eigenfrequencies. However, a change is noted in the rate of change in the curve at 5e4 s. This is attributed to the phase change between the movement of the primary and secondary mass, once the absorption frequency is crossed. The passive system has the second-highest dissipation energy and two clear saturation levels, namely 7.321 J and 15.635 J. These are attributed to the passage through the two resonance frequencies. In the absorption range, the energy remains nearly constant. The multiple model control has the third-highest dissipated energy. It shows only one range, in which the dissipated energy increases substantially, and reaches the saturation level of 9.556 J. The control strategy with the lowest dissipated energy is the slow frequency-based control. Similar to the passive system it shows two saturation levels, but with noticeably lower energy levels. The first saturation level has a value of 3.028 J, whereas the second has a level of 3.457 J. The low dissipation is attributed to the foci of the control strategy, namely vibration absorption and structural changes via sticking. Relative to the passive system, the slow frequency-based control and the multiple model control respectively dissipate 77.89 % and 38.88 % less energy, whereas the Skyhook Control dissipates 36.74 % more energy than its passive counterpart.

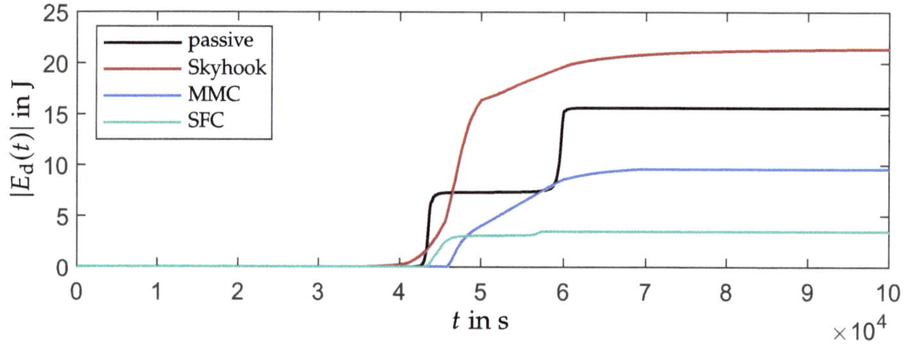

Figure 6.22: Comparison of the dissipated energy for the sweep excitation.

Figure 6.23 shows the dissipated energy for the application scenario. The curves in the application scenario have a similar progression. The largest increases in dissipated energy are observed during the overload and recovery phases. This is mainly due to the large oscillation amplitudes. Noticeable increases are also noted at the passages

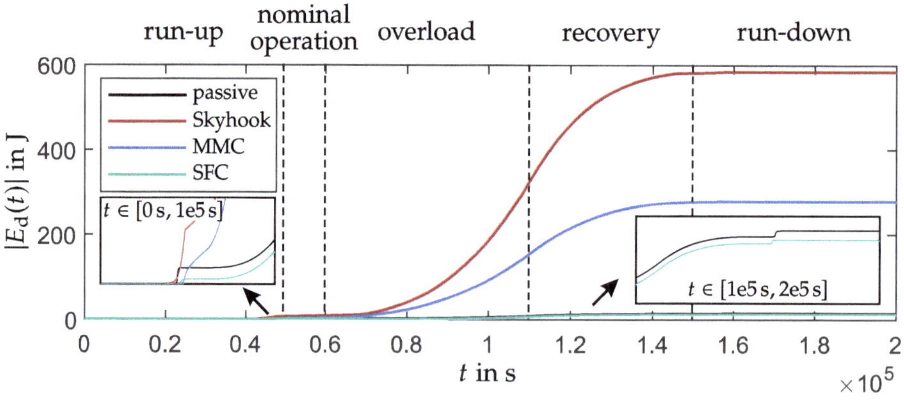

Figure 6.23: Comparison of the dissipated energy for the application scenario.

through resonances during the run-up and the run-down phases. The Skyhook Controls remains the strategy with the highest dissipated energy, namely 583.016 J at the end of the simulation. The multiple model control also has a high energy dissipation rate, and dissipates 278.295 J throughout the simulation. The passive system and the slow frequency-based control have similar results, namely 14.548 J and 12.592 J respectively. Thus, the slow frequency-based control dissipates the least amount of energy. The passive variant shows better results than the multiple model control since it relies on absorption during the overload and the recovery phases. Compared to the passive variant, the slow frequency-based control dissipates 13.44 % less energy. In contrast, the multiple model control and the Skyhook Control respectively dissipate 1812.90 % and 3907.45 % more energy relative to the passive system.

The minimal input energy is presented in Fig. 6.24 for the rectangular pulse. The curves show a similar progression compared to the dissipated energy, c.f. Fig. 6.21. In descending order of input energy, the Skyhook Control, the slow frequency-based control, and the multiple model control consume respectively 1.968 J, 0.003 J, and 6.489e-4 J. Although the slow frequency-based control dissipates less energy than the multiple model control, it uses more energy to vary the wedge angle and the prestress displacement. This is due to the larger oscillations of the relative coordinate $x_{\text{rel}} = x_2 - x_1$ over a longer time period. These oscillations lead to an increase in the input power by the prestress displacement. Apart from the Skyhook Control, it is noted that the energies used by the control strategies of this work are lower than the dissipated energy. Since fully active systems directly generate the necessary force to reduce oscillations, the dissipated energy represents the minimal energy required by a such systems. Therefore, this confirms the advantage of a semi-active system over an active system. Furthermore, it shows that the Skyhook Control Strategy is not an optimal strategy for the proposed base system. Compared to multiple model control, the input energy of slow frequency-based control and Skyhook Control are higher by a factor of 4.61 and 3,033.17, respectively.
Figure 6.25 shows the input energy of the control strategies for the sweep excitation. The form of the curves is the same as with the dissipated energy, however, the saturation levels are different, c.f Fig. 6.22. Once again, the Skyhook Control consumes the most energy, namely 16.521 J. The second highest energy usage is found in the slow frequency-based control with 0.907 J. Requiring only 0.688 J, the multiple model control shows the lowest energy consumption. A noticeable spike is noted in the input energy of the slow frequency at the passage of the first resonance frequency. The main cause for this energy spike and the strategy's higher energy consumption is the input power of the prestress displacement, which is caused by higher oscillation amplitudes of the relative coordinate during the transitions between the sticking and slipping systems.

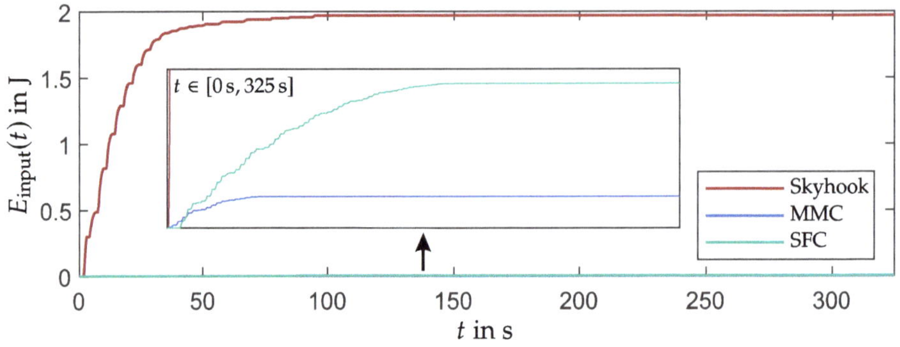

Figure 6.24: Comparison of the input energy for the rectangular pulse.

Compared to the multiple model control, the input energy of the slow frequency-based control and the Skyhook Control are higher by the factors 1.32 and 24.03.

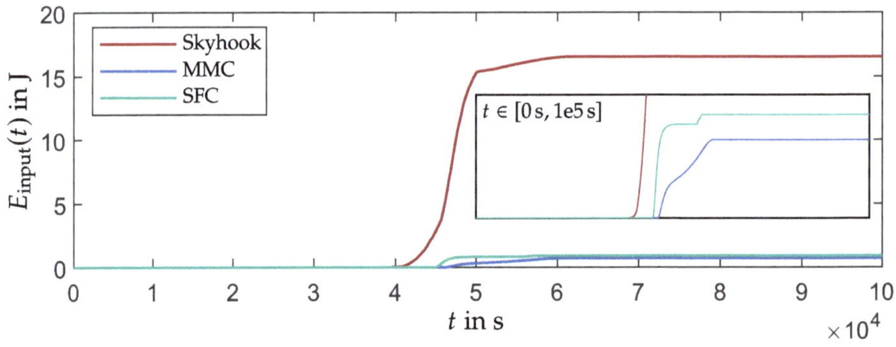

Figure 6.25: Comparison of the input energy for the sweep excitation.

The last observation regarding the damper energy is presented in Fig. 6.26 and considers the minimal input energy in the application scenario. The structure of the curves remains similar to those in Fig. 6.23. The Skyhook Control, the multiple model control, and the slow frequency-based control, respectively have a total input energy of 649.416 J, 32.235 J, and 0.702 J. Due to the use of absorption during the overload and recovery phase, where the oscillation amplitudes are largest, input energy of the slow frequency-based control is lower than that of the multiple model control. Compared to the slow frequency-based control, the input energy of the multiple model control and the Skyhook Control are respectively higher by a factor of 45.953 and 925.803.

With the energy assessments of the control strategies, conclusions regarding the influence of the strategies on the damper's service life and its energy costs are derived. The Skyhook Control shows the highest dissipated energy and the highest input energy. In all cases,

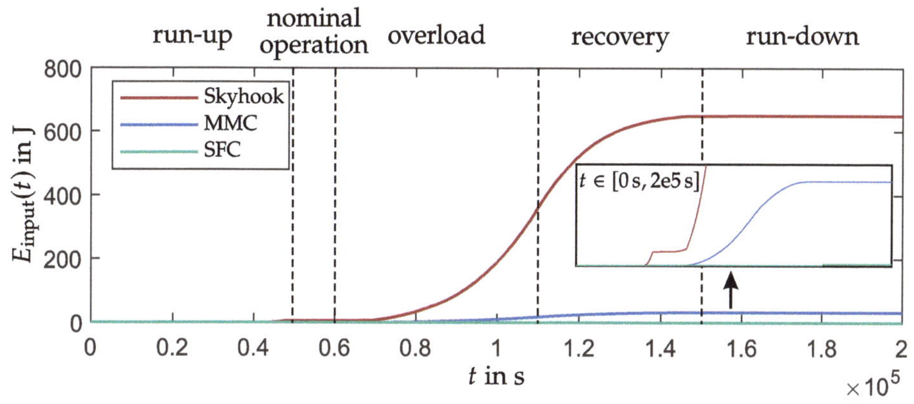

Figure 6.26: Comparison of the input energy for the application scenario.

the input energy was higher than the dissipated energy. Therefore, this strategy is not ideal for the base system since it leads to short life spans and high energy costs. The multiple model control shows a medium energy dissipation. This is coupled with a low energy requirement. The strategy, therefore, results in an acceptable service life with low energy costs, provided the design is self-locking. Finally, the slow frequency-based control provides a solution with low energy dissipation and low input energy. Therefore, the strategy results in a long service life with low energy consumption.

6.7 Control Strategy Assessment

This section presents an assessment of the control strategies. It condenses the results of the excitation simulations and evaluates the strategies' suitability for vibration reduction. First, the Skyhook Control is addressed. Subsequently, the multiple model control is evaluated. The section is closed with the assessment of the slow frequency-based control.

The Skyhook Control is a well-investigated control strategy, and it is taken as a reference for the developed strategies. It maximizes damping whenever possible and prevents exacerbating oscillations. The best performance of the control strategy is found in the response to the rectangular pulse. Apart from the ranges where the passive system utilizes absorption, the Skyhook Control shows an overall better response than the passive system. The disadvantages of the system are noted in the overload phase of the application scenario. The changes in the excitation amplitude lead to changes in the maximum input values. This extended input range proves counter-effective for vibration reduction. The strategy introduces more damping when absorption is the ideal damping mechanism, which leads to higher amplitudes. Energy-wise, the strategy has high energy dissipation and high energy consumption.

The adaptive multiple model control simulates three models in parallel and decides which is better suited for vibration reduction. The models are measured with a performance index, which takes the history of the vibration mitigation into account. Consequently, the model with the best performance index is chosen. The strategy is effective at mitigating transient excitations. However, as seen in Fig. 6.9, small oscillations remain. These oscillations are manageable since they are small and material damping present in practical applications helps mitigate such behavior. In the sweep response, the control strategy shows the second-best results. It has low amplitudes in comparison to the Skyhook Control and the passive system, however, it does not achieve absorption. The control strategy does not recognize this as the most effective vibration reduction mechanism. Other performance functions and parameters may offer better results. Furthermore, the control strategy is computationally costly since three models are simulated in parallel. However, the strategy has the advantage that it can adapt to unforeseen circumstances, e.g. polyharmonic and nonharmonic excitations. Finally, the strategy yields good results as long as the actuation models accurately represent the controlled system. For practical applications a high model accuracy and a corresponding identification are required. Regarding the energy investigations, this strategy results in moderate energy dissipation and low energy consumption.

The slow frequency-based control is designed for the stationary response of the system. To this end, it measures the position of the secondary mass, identifies the oscillation frequency, and with it determines the input parameters. The parameters are chosen to realize a targeted change between sticking and slipping. It has acceptable results in transient processes. It reduces the system's vibration, however, small oscillations remain. As with the multiple model control, these oscillations are manageable. Its sweep response is nearly optimal. The strategy avoids resonances and also achieves absorption. The advantage of the latter vibration reduction mechanism is noted in the overload phase of the application scenario. The vibrations remain smaller than the maximal amplitudes during the passage through resonance even though the excitation amplitude increases tenfold. This confirms the advantages of control strategies that are not solely dissipation-focused. It is also advantageous in slow-changing processes or in situations where the frequency characteristics of the process are known. In these cases, the frequency of the absorber is tuned to generate the best possible response. Furthermore, the quality of the strategy depends highly on the identification of the oscillation frequency. This can be problematic when more than one excitation frequency or excessive noise are involved. Energy-wise, the energy dissipated and its minimal input energy are low.

7 Conclusions and Future Work

This chapter concludes this work and gives a starting point for future investigations. First, the fulfillment of the thesis purposes and the sub-objectives stated in chapter 1 is assessed. The most important findings are reiterated. After the studies in this work are addressed, starting points for further investigations are proposed. These starting points are given according to the proposed sub-objectives.

The first sub-objective considered the design of passive dry friction dampers with piecewise-defined contact surfaces. An effective alternative to conventional dampers was suggested. The tuned wedge damper utilizes vibration damping in combination with absorption. However, it decouples these two vibration reduction mechanisms. Furthermore, the damper has the advantage of robustness, i.e. the scalability of the frequency response function with respect to the intensity of the excitation. Low amplitudes were found for high friction coefficients, however, such high values have a detrimental effect on the lifespan of the damper. Furthermore, in situations in which the material pairing is subject to additional restrictions, the increase of the friction coefficient is not a reliable choice. Low vibration amplitudes were reliably obtained via a simultaneous optimization of the system's tertiary stiffness and wedge angle.

The experimental validation of the theoretical investigations into passive systems with piecewise-defined contact surfaces is the focus of the second sub-objective. A prototype of the tuned wedge damper system was designed and tested with sweep excitations. The experiments validated the qualitative behavior concerning the variation of the tertiary stiffness, the friction coefficient, the prestress displacement, and the excitation force amplitude. The most important finding is the validation of the decoupling of damping and absorption. Some of the theoretical effects, however, were not observed in the experiments. For example, solutions with quasiperiodic or chaotic behavior were not found. Ranges with multiple solutions were also not verified. The sum of the experimental investigations offers the first proof of concept of the implementation of the tuned wedge damper as an practical vibration reduction mechanism.

The design of active dry friction dampers with piecewise-defined contact surfaces was considered in the third sub-objective. To this end, two strategies that do not solely focus on damping were proposed. Both control strategies show overall better results than a damping-driven strategy. The first strategy, the multiple model control, reduces vibrations by combining the three vibration reduction mechanisms damping, absorption, and shifting of structural resonances via sticking. It has a high adaptability degree, however, this comes with a high computational cost. The second strategy, the slow frequency-based control, changes its parameters depending on the oscillation frequency. It achieves the best results for a sweep excitation, namely the lowest maximal amplitudes while retaining the absorption frequency. Since both strategies do not focus on damping, they contribute to the enhancement of the life span of active friction dampers. Furthermore, they confirm that vibration reduction devices, that combine different vibration reduction mechanisms, achieve overall better results.

This thesis fulfills its main purpose, namely the design and validation of dry friction dampers that rely not solely on dissipation, but instead consider vibration absorption and stick-slip transitions to reduce vibrations. The passive investigations and experiments provide a damper that achieves this by effectively using piecewise-defined contact surfaces. Additionally, the investigations of active designs yield two active control strategies. The adaptive multiple model control unites the three vibration mechanisms, whereas the slow frequency-based control strategy does not rely on damping, but instead focuses on absorption and resonance avoidance with a targeted stick-slip transition. These investigations provide a starting point for additional investigations into mechanisms that combine the vibration reduction mechanisms instead of focusing on one single strategy.

Considering future work, the passive damping devices only took into account external excitations. The behavior of the tuned wedge damper with self-excited or parameter excited oscillations was not considered. It is important to assess the damper's effectiveness with different types of excitation. In addition, the damper's effectiveness in system with multiple degrees of freedom is of interest. To this end, the damper's design as well as its placement should be considered. Taking into account polynomial outer segments is an additional modification to consider. These segments can offer additional damping when it is most required, thus, leading to a more robust damper with respect to the excitation amplitude. Since the damper does not solely rely on damping, it saves energy. The application of the tuned wedge damper within a system force-flow should, therefore, be assessed. This placement could improve machine efficiency.

Extensions of the experimental investigations include more parameter variations and the quantitative reproduction of the experimental results. The reproduction requires a minimal model with damping and an identification of the damping present in the prototype. Another extension is the thorough investigation of the nonperiodic solutions and multiple solution ranges in an experimental manner. To this end, a test bench with less damping is required, e.g. a test bench with air bearings. A more practical design of the damper, which is closer to series production, is also of investigative value. This has the advantage that it can be used in more realistic situations, e.g. buildings, bridges, turbines, wheel suspensions, and drive trains.

The investigations into the active tuned wedge damper can also be expanded by taking into account different types of excitations, i.e. self-excited and parameter excited oscillations. Furthermore, the full vibration amplitude reduction can be exploited by applying the control strategies to a damper with optimized parameters. Additionally, the influence of different performance functions, parameter combinations, and additional models could improve the vibration reduction of the multiple model control. Additional scenarios should also be taken into account. For example, the damper can be placed in a drive train and different driving cycles can be tested. The ability to dampen vibrations as well as save energy in these applications should be assessed. To improve the approximation of the consumed energy, actuation models should be taken into account. Lastly, the experimental design of the tuned wedge damper, as well as the experimental validation of the control strategies, can offer a proof of concept for the practical implementation of the active tuned wedge damper.

Appendix

A Integration of Piecewise Terms of the Tuned Wedge Damper

This appendix handles the integration of the piecewise terms in the tuned wedge damper. The starting point for the integration is the tuned wedge damper force in its nondimensionalized and modified form

$$\begin{aligned}\tilde{f}_{\mathrm{TWD,lin}} =& a y y_x + b y_x + c y (y_x^2 + 1)\operatorname{sgn}(x_2' - x_1') + d (y_x^2 + 1)\operatorname{sgn}(x_2' - x_1') \\ &- 4a \tan^2 \alpha (x_2 - x_1)\,.\end{aligned} \tag{A.1}$$

The functions y and y_x are approximated with a zero order TAYLOR series in r. This is applicable since $r = \mathcal{O}(\varepsilon)$ and all higher order terms are irrelevant for the first order averaging method. This simplification results in $\Delta_2 = \Delta_1$ and

$$y \approx \begin{cases} -\tan \alpha (x_{\mathrm{rel}} + \Delta_1), & x_{\mathrm{rel}} < -\Delta_1 \\ 0, & |x_{\mathrm{rel}}| < \Delta_1 \\ \tan \alpha (x_{\mathrm{rel}} - \Delta_1), & \Delta_1 < x_{\mathrm{rel}} \end{cases} \quad \text{and} \quad y_x \approx \begin{cases} -\tan \alpha, & x_{\mathrm{rel}} < -\Delta_1 \\ 0, & |x_{\mathrm{rel}}| < \Delta_1 \\ \tan \alpha, & \Delta_1 < x_{\mathrm{rel}} \end{cases}. \tag{A.2}$$

In an exemplary manner, the terms $f_a = a y y_x$ are considered. However, the same procedure is applied for all the terms in $\tilde{f}_{\mathrm{TWD,lin}}$. The piecewise definition of y and y_x results in the piecewise defined force

$$f_a = \begin{cases} f_{a,1}, & x_{\mathrm{rel}} < -\Delta_1 \\ f_{a,2}, & |x_{\mathrm{rel}}| < \Delta_1 \\ f_{a,3}, & \Delta_1 < x_{\mathrm{rel}} \end{cases} = a \begin{cases} \tan^2 \alpha (x_{\mathrm{rel}} + \Delta_1), & x_{\mathrm{rel}} < -\Delta_1 \\ 0, & |x_{\mathrm{rel}}| < \Delta_1 \\ \tan^2 \alpha (x_{\mathrm{rel}} - \Delta_1), & \Delta_1 < x_{\mathrm{rel}} \end{cases}. \tag{A.3}$$

The transition points where the relative displacement x_{rel} surpasses the Δ_1 thresholds are identified. Since these terms are integrated over the fast oscillating phases, the transitions of x_{rel} are identified in dependence of φ_i. To this end, the VAN DER POL

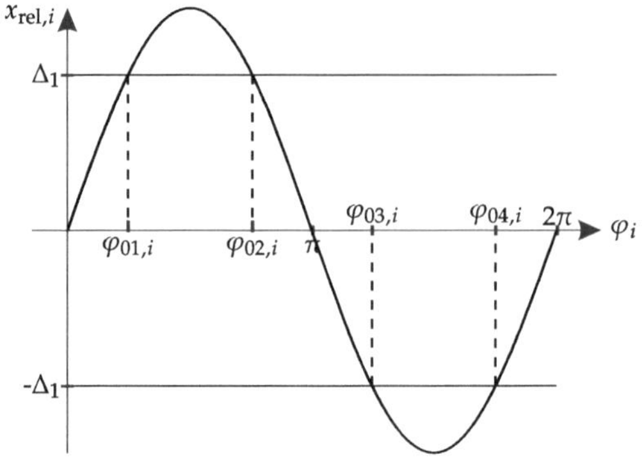

Figure A.1: Transition of the relative coordinate $x_{\text{rel}} = A_i(r_{2i} - r_{1i}) \sin \varphi_i$.

transformation and the modal decoupling is applied to the relative coordinate. In dependence of the considered eigenfrequency η_{0i}, this yields

$$x_{\text{rel,i}} = A_i(r_{2i} - r_{1i}) \sin \varphi_i \,. \tag{A.4}$$

The transitions of the relative coordinate are visualized in Fig. A.1. Using the symmetry of the sine function the angles φ_{0i} are identified as

$$\varphi_{01,i} = \arcsin\left(\frac{\Delta_1}{(r_{2i} - r_{1i})A_i}\right), \qquad \varphi_{02,i} = \pi - \varphi_{01,i}\,,$$
$$\varphi_{03,i} = \pi + \varphi_{01,i}\,, \qquad \text{and} \qquad \varphi_{04,i} = 2\pi - \varphi_{01,i}\,.$$

The force f_a is multiplied by $\sin \varphi_i$ or $\cos \varphi_i$. The average of these products over one period is divided into four integrals. The cosine product results in

$$\begin{aligned}
I_{a,\cos} &= \frac{1}{2\pi} \int_0^{2\pi} f_a \cos \varphi_i \mathrm{d}\varphi_i \\
&= \frac{1}{2\pi} \left(\int_{-\varphi_{01,i}}^{\varphi_{01,i}} f_{a,2} \cos \varphi_i \mathrm{d}\varphi_i + \int_{\varphi_{01,i}}^{\varphi_{02,i}} f_{a,3} \cos \varphi_i \mathrm{d}\varphi_i \right. \\
&\qquad \left. + \int_{\varphi_{02,i}}^{\varphi_{03,i}} f_{a,2} \cos \varphi_i \mathrm{d}\varphi_i + \int_{\varphi_{03,i}}^{\varphi_{04,i}} f_{a,1} \cos \varphi_i \mathrm{d}\varphi_i \right) \\
&= \frac{1}{2\pi} \left(\int_{\varphi_{01,i}}^{\varphi_{02,i}} \tan^2 \alpha (A_i(r_{2i} - r_{1i}) \sin \varphi_i + \Delta_1) \cos \varphi_i \mathrm{d}\varphi_i \right. \\
&\qquad \left. + \int_{\varphi_{03,i}}^{\varphi_{04,i}} \tan^2 \alpha (A_i(r_{2i} - r_{1i}) \sin \varphi_i - \Delta_1) \cos \varphi_i \mathrm{d}\varphi_i \right) = 0\,.
\end{aligned} \tag{A.5}$$

Analogously, the sine product yields

$$
\begin{aligned}
I_{a,\sin} &= \frac{1}{2\pi}\int_0^{2\pi} f_a \sin\varphi_i \mathrm{d}\varphi_i \\
&= \frac{1}{2\pi}\Bigg(\int_{\varphi_{01,i}}^{\varphi_{02,i}} \tan^2\alpha(A_i(r_{2i}-r_{1i})\sin\varphi_i+\Delta_1)\sin\varphi_i\mathrm{d}\varphi_i \\
&\qquad + \int_{\varphi_{03,i}}^{\varphi_{04,i}} \tan^2\alpha(A_i(r_{2i}-r_{1i})\sin\varphi_i-\Delta_1)\sin\varphi_i\mathrm{d}\varphi_i\Bigg) \\
&= \frac{4\tan^2\alpha}{A_i\pi}\left(\Delta_1\sqrt{(r_{1i}-r_{2i})^2A_i^2-\Delta_1^2}+A_i^2(r_{1i}-r_{2i})^2\arcsin\left(\frac{\Delta_1}{(r_{1i}-r_{2i})A_i}\right)\right).
\end{aligned}
\tag{A.6}
$$

The procedure described above is applied to all the terms in $\tilde{f}_{\mathrm{TWD,lin}}$. Together with the average of the excitation, these terms yield the averaged differential equations, i.e. Eqs. (4.29) and (4.30).

List of Figures

Bibliography

[1] Aida, T., S. Toda, N. Ogawa, and Y. Imada (1992): Vibration control of beams by beam-type dynamic vibration absorbers. *Journal of Engineering Mechanics* **118**(2), pp. 248–258.

[2] Alkhatib, R. and M. Golnaraghi (2003): Active structural vibration control: a review. *Shock and Vibration Digest* **35**(5), pp. 367–383.

[3] Alspaugh, D. (1978): Analysis of Coulomb friction vibration dampers. *Journal of Sound and Vibration* **57**(1), pp. 65–78.

[4] Andrew, C., J. Cockburn, and A. Waring (1967): Metal surfaces in contact under normal forces: some dynamic stiffness and damping characteristics. In: *Proceedings of the Institution of Mechanical Engineers, Conference Proceedings*. Vol. 182. SAGE Publications Sage UK: London, England, pp. 92–100.

[5] Argyris, J. H., G. Faust, M. Haase, and R. Friedrich (2015): *An exploration of dynamical systems and chaos: completely revised and enlarged second edition*. Berlin: Springer.

[6] Armstrong-Hélouvry, B., P. Dupont, and C. C. De Wit (1994): A survey of models, analysis tools and compensation methods for the control of machines with friction. *Automatica* **30**(7), pp. 1083–1138.

[7] Aronov, V., A. D'Souza, S. Kalpakjian, and I. Shareef (1984): Interactions among friction, wear, and system stiffness—Part 1: effect of normal load and system stiffness. *Journal of Tribology* **106**(1), pp. 54–58.

[8] Asami, T. and O. Nishihara (2003): Closed-form exact solution to H∞ optimization of dynamic vibration absorbers (application to different transfer functions and damping systems). *Journal of Vibration and Acoustics* **125**(3), pp. 398–405.

[9] Avriel, M. (2003): *Nonlinear programming: analysis and methods*. Mineola, New York: Dover Publications.

[10] Babitsky, V. I. (2013): *Theory of vibro-impact systems and applications*. Berlin, Heidelberg: Springer Science & Business Media.

[11] Bailey, T. and J. E. Hubbard Jr (1985): Distributed piezoelectric-polymer active vibration control of a cantilever beam. *Journal of Guidance, Control, and Dynamics* **8**(5), pp. 605–611.

[12] Bapat, C. and S. Sankar (1985): Multiunit impact damper—re-examined. *Journal of Sound and Vibration* **103**(4), pp. 457–469.

[13] Bender, E. and P. Remington (1974): The influence of rails on train noise. *Journal of Sound and Vibration* **37**(3), pp. 321–334.

[14] Berger, E. (2002): Friction modeling for dynamic system simulation. *Applied Mechanics Reviews* **55**(6), pp. 535–577.

[15] Bhushan, B. (1987): Magnetic head-media interface temperatures: part 1—analysis. *Journal of Tribology* **109** (1), pp. 243–251.

[16] Bhushan, B. (2013): *Introduction to tribology*. Chichester, West Sussex: John Wiley & Sons.

[17] Bielawa, R. (1978): An analytic study of the energy dissipation of turbomachinery bladed-disk assemblies due to inter-shroud segment rubbing. *Journal of Mechanical Design* **100** (1), pp. 223–228.

[18] Blok, H. (1940): Fundamental aspects of boundary friction. *Journal of Society of Automotive Engineers* **46** (1), p. 275.

[19] Bograd, S., P. Reuss, A. Schmidt, L. Gaul, and M. Mayer (2011): Modeling the dynamics of mechanical joints. *Mechanical Systems and Signal Processing* **25**(8), pp. 2801–2826.

[20] Bouckaert, S., A. Fernandez Pales, C. McGlade, U. Remme, and B. Wanner (2021): *Net Zero by 2050 A Roadmap for the Global Energy Sector*. Tech. rep. International Energy Agency.

[21] Bowden, F. P., F. P. Bowden, and D. Tabor (2001): *The friction and lubrication of solids Band I*. New York: Oxford University press.

[22] Bowden, F. P. and L. Leben (1939): The nature of sliding and the analysis of friction. *Proceedings of the Royal Society of London. Series A. Mathematical and Physical Sciences* **169**(938), pp. 371–391.

[23] Bowden, F. P. and D. Tabor (1964): *The Friction and Lubrication of Solids Part II*. New York: Oxford University Press.

[24] Brockley, C., R. Cameron, and A. Potter (1967): Friction-induced vibration. *Journal of Lubrication Technology* **89**(2), pp. 101–107.

[25] Brockley, C. and P. L. Ko (1970): Quasi-harmonic friction-induced vibration. *Journal of Tribology* **92**(4), pp. 550–556.

[26] Canudas De Wit, C., H. Olsson, K. J. Astrom, and P. Lischinsky (1995): A new model for control of systems with friction. *IEEE Transactions on automatic control* **40**(3), pp. 419–425.

[27] Chan, S. and I. Tuba (1971): A finite element method for contact problems of solid bodies—part II. Application to turbine blade fastenings. *International Journal of Mechanical Sciences* **13**(7), pp. 627–639.

[28] Childs, D. W. (1979): Rub-induced parametric excitation in rotors. *Journal of Mechanical Design* **101**(4), pp. 640–644.

[29] Constantinou, M. C. and I. G. Tadjbakhsh (1984): The optimum design of a base isolation system with frictional elements. *Earthquake Engineering & Structural Dynamics* **12**(2), pp. 203–214.

[30] Csaba, G. (1999): Modelling of a microslip friction damper subjected to translation and rotation. In: *ASME 1999 international gas turbine and aeroengine congress and exhibition.* American Society of Mechanical Engineers, 99-GT–149.

[31] Dahl, P. R. (1968): *A solid friction model.* Tech. rep. Aerospace Corp El Segundo Ca.

[32] Davis, C. L. and G. A. Lesieutre (2000): An actively tuned solid-state vibration absorber using capacitive shunting of piezoelectric stiffness. *Journal of Sound and Vibration* **232**(3), pp. 601–617.

[33] Den Hartog, J. P. (1985): *Mechanical vibrations.* New York: Courier Corporation.

[34] Den Hartog, J. (1930): Forced vibrations with combined viscous and coulomb damping. *The London, Edinburgh, and Dublin Philosophical Magazine and Journal of Science* **9**(59), pp. 801–817.

[35] Desplanques, Y. (2015): Amontons-Coulomb Friction Laws, A Review of the Original Manuscript. *Journal of Materials and Manufacturing* **8**(1), pp. 98–103.

[36] Detroux, T., G. Habib, L. Masset, and G. Kerschen (2015): Performance, robustness and sensitivity analysis of the nonlinear tuned vibration absorber. *Mechanical Systems and Signal Processing* **60**, pp. 799–809.

[37] Dowell, E. and H. Schwartz (1983): Forced response of a cantilever beam with a dry friction damper attached, Part I: Theory. *Journal of Sound and Vibration* **91**(2), pp. 255–267.

[38] Dowson, D. (1979): *History of Tribology.* London, New York: Addison-Wesley Longman Limited.

[39] Dupont, P., B. Armstrong, and V. Hayward (2000): Elasto-plastic friction model: contact compliance and stiction. In: *Proceedings of the 2000 American control conference.* Vol. 2. Institute of Electrical and Electronics Engineers, pp. 1072–1077.

[40] Dupont, P., V. Hayward, B. Armstrong, and F. Altpeter (2002): Single state elasto-plastic friction models. *IEEE Transactions on Automatic Control* **47**(5), pp. 787–792.

[41] Dupont, P., P. Kasturi, and A. Stokes (1997): Semi-active control of friction dampers. *Journal of Sound and Vibration* **202**(2), pp. 203–218.

[42] Dweib, A. and A. D'souza (1990): Self-excited vibrations induced by dry friction, part 2: Stability and limit-cycle analysis. *Journal of Sound and Vibration* **137**(2), pp. 177–190.

[43] Dyke, S., B. Spencer Jr, M. Sain, and J. Carlson (1998): An experimental study of MR dampers for seismic protection. *Smart Materials and Structures* **7**(5), pp. 693–703.

[44] Eckhaus, W. (1975): New approach to the asymptotic theory of nonlinear oscillations and wave-propagation. *Journal of Mathematical Analysis and Applications* **49**(3), pp. 575–611.

[45] Elias, S. and V. Matsagar (2017): Research developments in vibration control of structures using passive tuned mass dampers. *Annual Reviews in Control* **44**, pp. 129–156.

[46] Ewins, D. J. (2009): *Modal testing: theory, practice and application.* Baldock, Hertforshire, England: John Wiley & Sons.

[47] Farid, M. and O. Gendelman (2017): Response regimes in equivalent mechanical model of strongly nonlinear liquid sloshing. *International Journal of Non-Linear Mechanics* **94**, pp. 146–159.

[48] Feeny, B., A. s. Guran, N. Hinrichs, and K. Popp (1998): A historical review on dry friction and stick-slip phenomena. *Applied Mechanics Reviews* **51**(5), pp. 321–341.

[49] Fenz, D. M. and M. C. Constantinou (2006): Behaviour of the double concave friction pendulum bearing. *Earthquake Engineering & Structural Dynamics* **35**(11), pp. 1403–1424.

[50] Ferri, A. (1995): Friction damping and isolation systems. *Journal of Vibration and Acoustics* **117**(B), pp. 196–206.

[51] Fidlin, A. (2005): *Nonlinear oscillations in mechanical engineering.* Berlin, New York: Springer Science & Business Media.

[52] Fidlin, A., O. Drozdetskaya, and B. Waltersberger (2011): On the minimal model for the low frequency wobbling instability of friction discs. *European Journal of Mechanics-A/Solids* **30**(5), pp. 665–672.

[53] Fidlin, A. and N. Gafur (2017): On the dynamics of friction based tuned mass dampers. In: *Proceedings of the 2017 European nonlinear oscilations conference ENOC.* European Mechanics Society, p. ID87.

[54] Fidlin, A. and M. Lobos (2014): On the limiting of vibration amplitudes by a sequential friction-spring element. *Journal of Sound and Vibration* **333**(23), pp. 5970–5979.

[55] Fidlin, A. and W. Stamm (2009): On the radial dynamics of friction disks. *European Journal of Mechanics-A/Solids* **28**(3), pp. 526–534.

[56] Filiatrault, A., R. Tremblay, and R. Kar (2000): Performance evaluation of friction spring seismic damper. *Journal of Structural Engineering* **126**(4), pp. 491–499.

[57] Filippov, A. F. (2013): *Differential equations with discontinuous righthand sides: control systems*. Dordrecht, Netherlands: Springer Science & Business Media.

[58] Fosberry, R. A. C. and Z. Holubecki (1955): *An investigation of the cause and nature of brake squeal*. Lindley: Motor Industry Research Association.

[59] Frahm, H. (1911): *Devices for damping vibration of bodies*. US Patent No. 989,958.

[60] Fujino, Y., L. Sun, B. M. Pacheco, and P. Chaiseri (1992): Tuned liquid damper (TLD) for suppressing horizontal motion of structures. *Journal of Engineering Mechanics* **118**(10), pp. 2017–2030.

[61] Fujita, T. (1994): Application of hybrid mass damper with convertible active and passive modes using hydraulic actuator to high-rise building. In: *Proceedings of 1994 American Control Conference-ACC'94*. Vol. 1. Institute of Electrical and Electronics Engineers, pp. 1067–1072.

[62] Gafur, N. (2017): *Untersuchung der dynamischen Wirkung eines reibungsbasierten Tilgers mittels der Mittelwertbildung*. (unpublished). Masther thesis. Karlsruher Institut für Technologie.

[63] Gao, C. and D. Kuhlmann-Wilsdorf (1990): On stick-slip and the velocity dependence of friction at low speeds. *Journal of Tribology* **112**(2), pp. 354–360.

[64] Gardner, J. F. and J. P. Cusumano (1997): Dynamic models of friction wedge dampers. In: *Proceedings of the 1997 IEEE/ASME Joint Railroad Conference*. Institute of Electrical and Electronics Engineers, pp. 65–69.

[65] Gaul, L., H. Albrecht, and J. Wirnitzer (2004): Semi-active friction damping of large space truss structures. *Shock and Vibration* **11**(3-4), pp. 173–186.

[66] Gaul, L. and R. Nitsche (2001): The role of friction in mechanical joints. *Applied Mechanics Reviews* **54**(2), pp. 93–106.

[67] Godfrey, D. (1967): Vibration reduces metal to metal contact and causes an apparent reduction in friction. *ASLE Transactions* **10**(2), pp. 183–192.

[68] Greenwood, J. A. and J. Williamson (1966): Contact of nominally flat surfaces. *Proceedings of the Royal Society of London. Series A. Mathematical and Physical Sciences* **295**(1442), pp. 300–319.

[69] Griffin, J. and R. Labelle (1996): A rational method for optimizing shroud damping. In: *Turbo Expo: Power for Land, Sea, and Air*. Vol. 78767. American Society of Mechanical Engineers, 96-GT–402.

[70] Grosch, K. (1963): The relation between the friction and visco-elastic properties of rubber. *Proceedings of the Royal Society of London. Series A. Mathematical and Physical Sciences* **274**(1356), pp. 21–39.

[71] Gutierrez Soto, M. and H. Adeli (2013): Tuned mass dampers. *Archives of Computational Methods in Engineering* **20**(4), pp. 419–431.

[72] Habib, G., T. Detroux, R. Viguié, and G. Kerschen (2015): Nonlinear generalization of Den Hartog's equal-peak method. *Mechanical Systems and Signal Processing* **52**, pp. 17–28.

[73] Haessig Jr, D. A. and B. Friedland (1991): On the modeling and simulation of friction. *Journal of Dynamic Systems, Measurement, and Control* **113**(3), pp. 354–362.

[74] Hart, J. D., R. Sause, G. W. Ford, and D. G. Row (1992): Mitigation of wind-induced vibration of arctic pipeline systems. In: *Proceedings of the International Conference on Offshore Mechanics and Arctic Engineering*. American Society of Mechanical Engineers, pp. 169–169.

[75] Hausner, M., S. Lehmann, and M. Häßler (2011): *Reibungskupplung mit einer Kupplungsscheibe zur Übertragung von Drehmomenten*. DE 102010049929A1. Schaeffler Technologies GmbH.

[76] Hervé, B., J. Sinou, H. Mahé, and L. Jézéquel (2008): Analysis of friction-induced self-generated vibrations originated from mode-coupling in clutches. *International Journal of Pure and Applied Mathematics* **42**(3), pp. 369–375.

[77] Hetzler, H., D. Schwarzer, and W. Seemann (2007): Analytical investigation of steady-state stability and Hopf-bifurcations occurring in sliding friction oscillators with application to low-frequency disc brake noise. *Communications in Nonlinear Science and Numerical Simulation* **12**(1), pp. 83–99.

[78] Hetzler, H. (2009): On moving continua with contacts and sliding friction: Modeling, general properties and examples. *International Journal of Solids and Structures* **46**(13), pp. 2556–2570.

[79] Hisakado, T. (1974): Effect of surface roughness on contact between solid surfaces. *Wear* **28**(2), pp. 217–234.

[80] Hoffmann, N., M. Fischer, R. Allgaier, and L. Gaul (2002): A minimal model for studying properties of the mode-coupling type instability in friction induced oscillations. *Mechanics Research Communications* **29**(4), pp. 197–205.

[81] Housner, G., L. Bergman, T. Caughey, A. G. Chassiakos, R. O. Claus, S. F. Masri, R. E. Skelton, T. Soong, B. Spencer, and J. T. Yao (1997): Structural control: past, present, and future. *Journal of Engineering Mechanics* **123**(9), pp. 897–971.

[82] Ibrahim, R. (1994): Friction-induced vibration, chatter, squeal, and chaos—part I: mechanics of contact and friction. *Applied Mechanics Reviews* **47**(7), pp. 209–226.

[83] Ibrahim, R. (1994): Friction-induced vibration, chatter, squeal, and chaos—part II: dynamics and modeling. *Applied Mechanics Reviews* **47**(7), pp. 227–253.

[84] Ibrahim, R. and C. Pettit (2005): Uncertainties and dynamic problems of bolted joints and other fasteners. *Journal of Sound and Vibration* **279**(3-5), pp. 857–936.

[85] Inaudi, J. A. (1997): Modulated homogeneous friction: a semi-active damping strategy. *Earthquake Engineering & Structural Dynamics* **26**(3), pp. 361–376.

[86] Jo, H. and H. Yabuno (2010): Amplitude reduction of parametric resonance by dynamic vibration absorber based on quadratic nonlinear coupling. *Journal of Sound and Vibration* **329**(11), pp. 2205–2217.

[87] Kaiser, A. B., J. Cusumano, and J. Gardner (2002): Modeling and dynamics of friction wedge dampers in railroad freight trucks. *Vehicle System Dynamics* **38**(1), pp. 55–82.

[88] Kandasamy, R., F. Cui, N. Townsend, C. C. Foo, J. Guo, A. Shenoi, and Y. Xiong (2016): A review of vibration control methods for marine offshore structures. *Ocean Engineering* **127**, pp. 279–297.

[89] Kapelke, S. (2019): *Zur Beeinflussung reibungsbehafteter Systeme mithilfe überlagerter Schwingungen.* (Dissertation). Karlsruhe: KIT Scientific Publishing.

[90] Karnopp, D. (1985): Computer simulation of stick-slip friction in mechanical dynamic systems. *Journal of Dynamic Systems, Measurement, and Control* **107**(1), pp. 100–107.

[91] Karnopp, D., M. J. Crosby, and R. Harwood (1974): Vibration control using semi-active force generators. *Journal of Manufacturing Science and Engineering* **96**(2), pp. 619–626.

[92] Keer, L., N. Ahmadi, and T. Mura (1984): Tangential loading of elastic bodies in contact. *Computers & Structures* **19**(1-2), pp. 93–101.

[93] Kelly, J. M. (1986): A seismic base isolation: review and bibliography. *Soil Dynamics and Earthquake Engineering* **5**(4), pp. 202–216.

[94] Khoo, H.-H., C. Clifton, J. Butterworth, G. MacRae, S. Gledhill, and G. Sidwell (2012): Development of the self-centering Sliding Hinge Joint with friction ring springs. *Journal of Constructional Steel Research* **78**, pp. 201–211.

[95] Kim, S.-M., S. Wang, and M. J. Brennan (2011): Optimal and robust modal control of a flexible structure using an active dynamic vibration absorber. *Smart Materials and Structures* **20**(4), p. 045003.

[96] Kinkaid, N., O. M. O'Reilly, and P. Papadopoulos (2003): Automotive disc brake squeal. *Journal of Sound and Vibration* **267**(1), pp. 105–166.

[97] Ko, P. L. and C. Brockley (1970): The measurement of friction and friction-induced vibration. *Journal of Tribology* **92**(4), pp. 543–549.

[98] Kovalev, R., N. Lysikov, G. Mikheev, D. Pogorelov, V. Simonov, V. Yazykov, S. Zakharov, I. Zharov, I. Goryacheva, S. Soshenkov, et al. (2009): Freight car models and their computer-aided dynamic analysis. *Multibody System Dynamics* **22**(4), pp. 399–423.

[99] Krack, M., L. A. Bergman, and A. F. Vakakis (2016): On the efficacy of friction damping in the presence of nonlinear modal interactions. *Journal of Sound and Vibration* **370**, pp. 209–220.

[100] Krack, M. and J. Gross (2019): *Harmonic balance for nonlinear vibration problems.* Cham, Switzerland: Springer.

[101] Krauter, A. (1981): Generation of squeal/chatter in water-lubricated elastomeric bearings. *Journal of Tribology* **103**(3), pp. 406–412.

[102] Krylov, N. M. and N. N. Bogoliubov (2016): *Introduction to Non-Linear Mechanics.* Princeton, New Jersey: Princeton University Press.

[103] Kuhlmann-Wilsdorf, D. (1987): Temperatures at interfacial contact spots: dependence on velocity and on role reversal of two materials in sliding contact. *Journal of Tribology* **109**(2), pp. 321–329.

[104] Laffranchi, M., L. Chen, N. Kashiri, J. Lee, N. G. Tsagarakis, and D. G. Caldwell (2014): Development and control of a series elastic actuator equipped with a semi active friction damper for human friendly robots. *Robotics and Autonomous Systems* **62**(12), pp. 1827–1836.

[105] Lane, J. S. and A. A. Ferri (1992): Optimal control of a semi-active, frictionally damped joint. In: *American Control Conference.* Institute of Electrical and Electronics Engineers, pp. 2754–2759.

[106] Lee-Glauser, G. J., G. Ahmadi, and L. G. Horta (1997): Integrated passive/active vibration absorber for multistory buildings. *Journal of Structural Engineering* **123**(4), pp. 499–504.

[107] Leine, R. I., D. Van Campen, and B. Van de Vrande (2000): Bifurcations in nonlinear discontinuous systems. *Nonlinear Dynamics* **23**(2), pp. 105–164.

[108] Lin, G.-L., C.-C. Lin, L.-Y. Lu, and Y.-B. Ho (2012): Experimental verification of seismic vibration control using a semi-active friction tuned mass damper. *Earthquake Engineering & Structural Dynamics* **41**(4), pp. 813–830.

[109] Liu, Y., T. Waters, and M. Brennan (2005): A comparison of semi-active damping control strategies for vibration isolation of harmonic disturbances. *Journal of Sound and Vibration* **280**(1-2), pp. 21–39.

[110] Love, J. and M. Tait (2015): Multiple tuned liquid dampers for efficient and robust structural control. *Journal of Structural Engineering* **141**(12), p. 04015045.

[111] Lu, L.-Y. (2004): Predictive control of seismic structures with semi-active friction dampers. *Earthquake Engineering & Structural Dynamics* **33**(5), pp. 647–668.

[112] Lu, Z., Z. Wang, S. F. Masri, and X. Lu (2018): Particle impact dampers: Past, present, and future. *Structural Control and Health Monitoring* **25**(1), e2058.

[113] Lu, Z., Z. Wang, Y. Zhou, and X. Lu (2018): Nonlinear dissipative devices in structural vibration control: A review. *Journal of Sound and Vibration* **423**, pp. 18–49.

[114] Ma, X., P. K. Wong, and J. Zhao (2019): Practical multi-objective control for automotive semi-active suspension system with nonlinear hydraulic adjustable damper. *Mechanical Systems and Signal Processing* **117**, pp. 667–688.

[115] MAISSEN, V. (1993): Festkoerperreibung-Reibungszahlen verschiedener Werkstoffe. *Schweizer Ingenieur und Architekt* **111**, pp. 25–29.

[116] MAKSIMOV, I. and A. RAKHMANOV (1987): The thermofrictional oscillations under sliding. *Physics Letters A* **121**(8-9), pp. 399–402.

[117] MARHADI, K. S. and V. K. KINRA (2005): Particle impact damping: effect of mass ratio, material, and shape. *Journal of Sound and Vibration* **283**(1-2), pp. 433–448.

[118] MARTINEZ-RODRIGO, M. and M. ROMERO (2003): An optimum retrofit strategy for moment resisting frames with nonlinear viscous dampers for seismic applications. *Engineering Structures* **25**(7), pp. 913–925.

[119] MARTINS, J., J. ODEN, and F. SIMOES (1990): A study of static and kinetic friction. *International Journal of Engineering Science* **28**(1), pp. 29–92.

[120] MARX, B. and W. VOGT (2010): *Dynamische Systeme: Theorie und Numerik*. Heidelberg: Springer-Verlag.

[121] MASRI, S. (1970): General motion of impact dampers. *Journal of the Acoustical Society of America* **47**(1B), pp. 229–237.

[122] MEGUID, S., P. KANTH, and A. CZEKANSKI (2000): Finite element analysis of fir-tree region in turbine discs. *Finite Elements in Analysis and Design* **35**(4), pp. 305–317.

[123] MINDLIN, R. D. (1949): Compliance of elastic bodies in contact. *Journal of Applied Mechanics* **16**(3), pp. 259–268.

[124] MITROPOLSKY, I. A. (1967): Averaging method in non-linear mechanics. *International Journal of Non-Linear Mechanics* **2**(1), pp. 69–96.

[125] MODI, V. and S. MUNSHI (1998): An efficient liquid sloshing damper for vibration control. *Journal of Fluids and Structures* **12**(8), pp. 1055–1071.

[126] MOON, F. C. and T. KALMAR-NAGY (2001): Nonlinear models for complex dynamics in cutting materials. *Philosophical Transactions of the Royal Society of London. Series A: Mathematical, Physical and Engineering Sciences* **359**(1781), pp. 695–711.

[127] MORGEN, B. and Y. KURAMA (2004): A friction damper for post-tensioned precast concrete beam-to-column joints. In: *13th World Conference on Earthquake Engineering*, p. 3189.

[128] MORSE, A. S. (1996): Supervisory control of families of linear set-point controllers-part i. exact matching. *IEEE Transactions on Automatic Control* **41**(10), pp. 1413–1431.

[129] NARENDRA, K. S. and J. BALAKRISHNAN (1997): Adaptive control using multiple models. *IEEE Transactions on Automatic Control* **42**(2), pp. 171–187.

[130] NEWCOMB, T. (1960): Temperatures reached in disc brakes. *Journal of Mechanical Engineering Science* **2**(3), pp. 167–177.

[131] NIMS, D. K., P. J. RICHTER, and R. E. BACHMAN (1993): The use of the energy dissipating restraint for seismic hazard mitigation. *Earthquake Spectra* **9**(3), pp. 467–489.

[132] Oden, J. and J. Martins (1985): Models and computational methods for dynamic friction phenomena. *Computer Methods in Applied Mechanics and Engineering* **52**(1-3), pp. 527–634.

[133] Pall, A. S., C. Marsh, et al. (1982): Response of friction damped braced frames. *Journal of Structural Engineering* **108**(9), pp. 1313–1323.

[134] Panning, L., W. Sextro, and K. Popp (2000): Optimization of interblade friction damper design. In: *Turbo Expo: Power for Land, Sea, and Air*. Vol. 78576. American Society of Mechanical Engineers, V004T03A068.

[135] Panossian, H. (1992): Structural damping enhancement via non-obstructive particle damping technique. *Journal of Vibration and Acoustics* **114**(1), pp. 101–105.

[136] Papanikos, P. and S. Meguid (1994): Theoretical and Experimental Studies of Fretting-Initiated Fatigue Failure of Aeroengine Compressor Discs. *Fatigue & Fracture of Engineering Materials & Structures* **17**(5), pp. 539–550.

[137] Parker, T. S. and L. Chua (2012): *Practical numerical algorithms for chaotic systems*. New York, Berlin, Heidelberg: Springer Science & Business Media.

[138] Plint, A. and M. Plint (1985): A new technique for the investigation of stick-slip. *Tribology International* **18**(4), pp. 247–249.

[139] Popp, K. and P. Stelter (1990): Nonlinear oscillations of structures induced by dry friction. In: *Nonlinear dynamics in engineering systems*. Springer, pp. 233–240.

[140] Popp, K., L. Panning, and W. Sextro (2003): Vibration damping by friction forces: theory and applications. *Journal of Vibration and Control* **9**(3-4), pp. 419–448.

[141] Popp, K. and P. Stelter (1990): Stick-slip vibrations and chaos. *Philosophical Transactions: Physical Sciences and Engineering* **332**(1624), pp. 89–105.

[142] Rabinowicz, E. (1951): The nature of the static and kinetic coefficients of friction. *Journal of Applied Physics* **22**(11), pp. 1373–1379.

[143] Remington, P. (1987): Wheel/rail squeal and impact noise: What do we know? What don't we know? Where do we go from here? *Journal of Sound and Vibration* **116**(2), pp. 339–353.

[144] Ricciardelli, F. and B. J. Vickery (1999): Tuned vibration absorbers with dry friction damping. *Earthquake Engineering & Structural Dynamics* **28**(7), pp. 707–723.

[145] Rizvi, A., C. W. Smith, R. Rajasekaran, and K. E. Evans (2016): Dynamics of dry friction damping in gas turbines: literature survey. *Journal of Vibration and Control* **22**(1), pp. 296–305.

[146] Rüdinger, F. (2006): Optimal vibration absorber with nonlinear viscous power law damping and white noise excitation. *Journal of Engineering Mechanics* **132**(1), pp. 46–53.

[147] Ruzicka, J. E. and T. F. Derby (1971): *Influence of damping in vibration isolation*. Washington: Shock and Vibration Information Center.

[148] SAKAMOTO, T. (1985): Normal displacement of the sliding body in a stick-slip friction process. In: *Proceedings of the JSLE International Tribology Conference*. Vol. 1. JSLE, pp. 141–146.

[149] SAMPSON, J., F. MORGAN, D. REED, and M. MUSKAT (1943): Studies in lubrication: XII. Friction behavior during the slip portion of the stick-slip process. *Journal of Applied Physics* **14**(12), pp. 689–700.

[150] SANDERS, J. A., F. VERHULST, and J. MURDOCK (2007): *Averaging methods in nonlinear dynamical systems*. New York: Springer.

[151] SARKAR, A. D. (1980): *Friction and wear*. London, New York: Academic Press.

[152] SASTRY, S. and M. BODSON (2011): *Adaptive control: stability, convergence and robustness*. Mineola, New York: Dover Publications.

[153] SCHNEIDER, E., K. POPP, and H. IRRETIER (1988): Noise generation in railway wheels due to rail-wheel contact forces. *Journal of Sound and Vibration* **120**(2), pp. 227–244.

[154] SCHUBERT, D. W. and J. E. RUZICKA (1969): Theoretical and experimental investigation of electrohydraulic vibration isolation systems. *Journal of Manufacturing Science and Engineering* **91**(4), pp. 981–990.

[155] SHAW, S. (1986): On the dynamic response of a system with dry friction. *Journal of Sound and Vibration* **108**(2), pp. 305–325.

[156] SMITH, R. E. and E. L. LUM (1968): Linear optimal theory applied to active structural bending control. *Journal of Aircraft* **5**(5), pp. 479–485.

[157] SMITH, R. and C. H. PAN (1975): *An Exploratory Laboratory Simulation of Stick-Slip Induced Vibration of Water-Lubricated Compliant-Layer Bearings*. Tech. rep. Shaker Research Corp Ballston Lake NY.

[158] SNOWDON, J. C. (1968): *Vibration and shock in damped mechanical systems*. Ann Arbor, Michigan: J. Wiley.

[159] SOKOLOV, I. and V. BABITSKY (2001): Phase control of self-sustained vibration. *Journal of Sound and Vibration* **248**(4), pp. 725–744.

[160] SPURR, R. T. (1961): A theory of brake squeal. *Proceedings of the Institution of Mechanical Engineers: Automobile Division* **15**(1), pp. 33–52.

[161] STACHOWIAK, G. W. (2006): *Wear: materials, mechanisms and practice*. Chichester, West Sussex: John Wiley & Sons.

[162] STAROSVETSKY, Y. and O. GENDELMAN (2009): Vibration absorption in systems with a nonlinear energy sink: nonlinear damping. *Journal of Sound and Vibration* **324**(3-5), pp. 916–939.

[163] STRIBECK, R. (1902): Die wesentlichen Eigenschaften der Gieit-und Rollenlager– the key qualities of sliding and roller bearings. *Zeitschrift des Vereines deutscher Ingenieure* **46**(38), pp. 1342–1348.

[164] SUN, J., M. R. JOLLY, and M. NORRIS (1995): Passive, adaptive and active tuned vibration absorbers—a survey. *Journal of Vibration and Acoustics* **117**(B), pp. 234–242.

[165] Sun, Y. and M. Thomas (2011): Control of torsional rotor vibrations using an electrorheological fluid dynamic absorber. *Journal of Vibration and Control* **17**(8), pp. 1253–1264.

[166] Symans, M. and M. Constantinou (1998): Passive fluid viscous damping systems for seismic energy dissipation. *ISET Journal of Earthquake Technology* **35**(4), pp. 185–206.

[167] Tabor, D. (1981): Friction—the present state of our understanding. *Journal of Tribology* **103**(2), pp. 169–176.

[168] Thomsen, J. J. and A. Fidlin (2003): Analytical approximations for stick–slip vibration amplitudes. *International Journal of Non-Linear Mechanics* **38**(3), pp. 389–403.

[169] Tlusty, J. (1985): Machine dynamics. In: *Handbook of high-speed machining technology*. Springer, pp. 48–153.

[170] Tobias, S. A. (1965): *Machine-tool vibration*. London: J. Wiley.

[171] Tomlinson, G. (1979): Force distortion in resonance testing of structures with electro-dynamic vibration exciters. *Journal of Sound and Vibration* **63**(3), pp. 337–350.

[172] Tubaldi, E., M. Barbato, and A. Dall'Asta (2014): Performance-based seismic risk assessment for buildings equipped with linear and nonlinear viscous dampers. *Engineering Structures* **78**, pp. 90–99.

[173] Vakakis, A. F. (2001): Inducing passive nonlinear energy sinks in vibrating systems. *Journal of Vibration and Acoustics* **123**(3), pp. 324–332.

[174] Vinogradov, G., I. Korepova, and Y. Y. Podolsky (1967): Steel-to-steel friction over a very wide range of sliding speeds. *Wear* **10**(5), pp. 338–352.

[175] von Wagner, U., D. Hochlenert, and P. Hagedorn (2007): Minimal models for disk brake squeal. *Journal of Sound and Vibration* **302**(3), pp. 527–539.

[176] Wagg, D. J. (2021): A review of the mechanical inerter: historical context, physical realisations and nonlinear applications. *Nonlinear Dynamics* **104**, pp. 13–34.

[177] Warn, G. P. and K. L. Ryan (2012): A review of seismic isolation for buildings: historical development and research needs. *Buildings* **2**(3), pp. 300–325.

[178] Weber, F. (2014): Semi-active vibration absorber based on real-time controlled MR damper. *Mechanical Systems and Signal Processing* **46**(2), pp. 272–288.

[179] Whitehouse, D. J. and J. Archard (1970): The properties of random surfaces of significance in their contact. *Proceedings of the Royal Society of London. A. Mathematical and Physical Sciences* **316**(1524), pp. 97–121.

[180] Wittmann, C., A. Tikhomolov, and M. Häßler (2011): Moderne Kupplungskonzepte für Fahrzeuge. In: *VDI-Berichte Nr. 2139 Kupplungen und Kupplungssysteme in Antrieben*. Verein Deutscher Ingenieure, pp. 35–47.

[181] WSJ Custom Studios and Emerson (2016): *Unlocking performance: how manufacturers can achieve top quartile performance*. url: `https://partners.wsj.com/`

emerson/unlocking-performance/how-manufacturers-can-achieve-top-quartile-performance/ (visited on 03/18/2022).

[182] Wu, J., Y. Xie, D. Zhang, and M. Zhang (2012): Experimental Friction Damping Characteristic of a Steam Turbine Blade Coupled by Shroud and Snubber at Standstill Set-Up. In: *Turbo Expo: Power for Land, Sea, and Air*. Vol. 44731. American Society of Mechanical Engineers, pp. 1315–1323.

[183] Wu, Q., C. Cole, M. Spiryagin, and Y.Q. Sun (2014): A review of dynamics modelling of friction wedge suspensions. *Vehicle System Dynamics* **52**(11), pp. 1389–1415.

Published Work

[184] Aramendiz, J. and A. Fidlin (2020): Analyse von Tilgerkonfigurationen mit trockenen Reibungselementen. *Forschung im Ingenieurwesen* **84**, pp. 179–189.

[185] Aramendiz, J., A. Fidlin, and E. Baranowski (2019): On the Dynamics of a Prestressed Sliding Wedge Damper. *PAMM* **19**(1), e201900326.

[186] Aramendiz, J., A. Fidlin, and K. Lei (2021): Investigations on amplitude adaptive sequential friction-spring dampers. *ZAMM-Journal of Applied Mathematics and Mechanics/Zeitschrift für Angewandte Mathematik und Mechanik* **101**(7), e201800293.

[187] Fidlin, A. and J. Aramendiz (2019): Study on the dynamics of a lock-up mass damper: asymptotic analysis and application limits. *Nonlinear Dynamics* **97**(3), pp. 1867–1875.

[188] Tan, A. S., J. Aramendiz, K. H. Ross, T. Sattel, and A. Fidlin (2019): Comparative study between dry friction and electrorheological fluid switches for Tuned Vibration Absorbers. *Journal of Sound and Vibration* **460**, p. 114874.

[189] Yüzbasioglu, T., J. Aramendiz, and A. Fidlin (2020): On the numerical simulations of amplitude-adaptive impact dampers. *Journal of Sound and Vibration* **468**, p. 115023.

Supervised Theses

[190] Baranowski, E. (2018): *Untersuchung der Dynamik eines reibungsbasierten Tilgers mit Keilen und Vorspannung*. (unpublished). Master thesis. Karlsruher Institut für Technologie.

[191] Lei, K. (2017): *Limitation of oscillations by the means of add-on sequential elastic stick-slip elements*. (unpublished). Master thesis. Karlsruher Institut für Technologie.

[192] Mayer, T. (2020): *Untersuchung der Dynamik eines vorgespannten trockenen Reibungstilgers mit glatter Geometrie*. (unpublished). Master thesis. Karlsruher Institut für Technologie.

[193] Yüzbasioglu, T. (2018): *Numerische Untersuchung von Stoßtilgern*. (unpublished). Master thesis. Karlsruher Institut für Technologie.

KARLSRUHER INSTITUT FÜR TECHNOLOGIE (KIT)
SCHRIFTENREIHE DES INSTITUTS FÜR TECHNISCHE MECHANIK (ITM)

ISSN 1614-3914

Band 1 **Marcus Simon**
Zur Stabilität dynamischer Systeme mit stochastischer Anregung.
ISBN 3-937300-13-9

Band 2 **Clemens Reitze**
Closed Loop, Entwicklungsplattform für mechatronische Fahrdynamikregelsysteme.
ISBN 3-937300-19-8

Band 3 **Martin Georg Cichon**
Zum Einfluß stochastischer Anregungen auf mechanische Systeme.
ISBN 3-86644-003-0

Band 4 **Rainer Keppler**
Zur Modellierung und Simulation von Mehrkörpersystemen unter Berücksichtigung von Greifkontakt bei Robotern.
ISBN 978-3-86644-092-0

Band 5 **Bernd Waltersberger**
Strukturdynamik mit ein- und zweiseitigen Bindungen aufgrund reibungsbehafteter Kontakte.
ISBN 978-3-86644-153-8

Band 6 **Rüdiger Benz**
Fahrzeugsimulation zur Zuverlässigkeitsabsicherung von karosseriefesten Kfz-Komponenten.
ISBN 978-3-86644-197-2

Band 7 **Pierre Barthels**
Zur Modellierung, dynamischen Simulation und Schwingungsunterdrückung bei nichtglatten, zeitvarianten Balkensystemen.
ISBN 978-3-86644-217-7

Band 8 **Hartmut Hetzler**
Zur Stabilität von Systemen bewegter Kontinua mit Reibkontakten am Beispiel des Bremsenquietschens.
ISBN 978-3-86644-229-0

Band 9 **Frank Dienerowitz**
Der Helixaktor – Zum Konzept eines vorverwundenen Biegeaktors.
ISBN 978-3-86644-232-0

Band 10 **Christian Rudolf**
Piezoelektrische Self-sensing-Aktoren zur Korrektur statischer Verlagerungen.
ISBN 978-3-86644-267-2

Band 11 **Günther Stelzner**
Zur Modellierung und Simulation biomechanischer Mehrkörpersysteme.
ISBN 978-3-86644-340-2

Band 12 **Christian Wetzel**
Zur probabilistischen Betrachtung von Schienen- und Kraftfahrzeugsystemen unter zufälliger Windanregung.
ISBN 978-3-86644-444-7

Band 13 **Wolfgang Stamm**
Modellierung und Simulation von Mehrkörpersystemen mit flächigen Reibkontakten.
ISBN 978-3-86644-605-2

Band 14 **Felix Fritz**
Modellierung von Wälzlagern als generische Maschinenelemente einer Mehrkörpersimulation.
ISBN 978-3-86644-667-0

Band 15 **Aydin Boyaci**
Zum Stabilitäts- und Bifurkationsverhalten hochtouriger Rotoren in Gleitlagern.
ISBN 978-3-86644-780-6

Band 16 **Rugerri Toni Liong**
Application of the cohesive zone model to the analysis of rotors with a transverse crack.
ISBN 978-3-86644-791-2

Band 17 **Ulrich Bittner**
Strukturakustische Optimierung von Axialkolbeneinheiten. Modellbildung, Validierung und Topologieoptimierung.
ISBN 978-3-86644-938-1

Band 18 **Alexander Karmazin**
Time-efficient Simulation of Surface-excited Guided Lamb Wave Propagation in Composites.
ISBN 978-3-86644-935-0

Band 19 **Heike Vogt**
Zum Einfluss von Fahrzeug- und Straßenparametern auf die Ausbildung von Straßenunebenheiten.
ISBN 978-3-7315-0023-0

Band 20 **Laurent Ineichen**
Konzeptvergleich zur Bekämpfung der Torsionsschwingungen im Antriebsstrang eines Kraftfahrzeugs.
ISBN 978-3-7315-0030-8

Band 21 **Sietze van Buuren**
Modeling and simulation of porous journal bearings in multibody systems.
ISBN 978-3-7315-0084-1

Band 22 **Dominik Kern**
Neuartige Drehgelenke für reibungsarme Mechanismen.
ISBN 978-3-7315-0103-9

Band 23 **Nicole Gaus**
Zur Ermittlung eines stochastischen Reibwerts und dessen Einfluss auf reibungserregte Schwingungen.
ISBN 978-3-7315-0118-3

Band 24 **Fabian Bauer**
Optimierung der Energieeffizienz zweibeiniger Roboter durch elastische Kopplungen.
ISBN 978-3-7315-0256-2

Band 25 **Benedikt Wiegert**
Nichtlineare Schwingungen von Systemen
mit elastohydrodynamischen Linienkontakten.
ISBN 978-3-7315-0350-7

Band 26 **Arsenty Tikhomolov**
Analytische, numerische und messtechnische
Untersuchung der Dynamik von Fahrzeugkupplungen
am Beispiel des Trennproblems.
ISBN 978-3-7315-0362-0

Band 27 **Daniel Maier**
On the Use of Model Order Reduction Techniques
for the Elastohydrodynamic Contact Problem.
ISBN 978-3-7315-0369-9

Band 28 **Xiaoyu Zhang**
Crosswind stability of vehicles under nonstationary
wind excitation.
ISBN 978-3-7315-0376-7

Band 29 **Jens Deppler**
Ein Beitrag zur viskoelastischen Modellierung nichtholonomer
Bindungsgleichungen.
ISBN 978-3-7315-0548-8

Band 30 **Georg Jehle**
Zur Modellbildung und Simulation reibungserregter
Schwingungen in Pkw-Schaltgetrieben.
ISBN 978-3-7315-0668-3

Band 31 **Joachim Klima**
Lubricant transport towards tribocontact in capillary
surface structures.
ISBN 978-3-7315-0814-4

Band 32 **Gábor Kenderi**
Nonparametric identification of nonlinear dynamic systems.
ISBN 978-3-7315-0834-2

Band 33 **Jan Henrik Schmidt**
An efficient solution procedure for elastohydrodynamic contact problems considering structural dynamics.
ISBN 978-3-7315-0872-4

Band 34 **Ulrich Johannes Römer**
Über den Einfluss der Fußgeometrie auf die Energieeffizienz beim zweibeinigen Gehen.
ISBN 978-3-7315-0887-8

Band 35 **Simon Kapelke**
Zur Beeinflussung reibungsbehafteter Systeme mithilfe überlagerter Schwingungen.
ISBN 978-3-7315-0905-9

Band 36 **Jens Burgert**
On direct and inverse problems related to longitudinal impact of non-uniform elastic rods.
ISBN 978-3-7315-1087-1

Band 37 **Tim Leister**
Dynamics of Rotors on Refrigerant-Lubricated Gas Foil Bearings.
ISBN 978-3-7315-1112-0

Band 38 **Simon Schröders**
Ein Beitrag zur asymptotischen Analyse der hydraulischen Ventilsteuerungen.
ISBN 978-3-7315-1221-9

Band 39 **Lukas Julian Oestringer**
Thermomechanischer Kontakt und stochastische Dynamik gleitender Körper mit rauen Oberflächen.
ISBN 978-3-7315-1297-4

Band 40 **Jimmy Alberto Aramendiz Fuentes**
On the Influence of Piecewise Defined Contact Geometries on Friction Dampers.
ISBN 978-3-7315-1267-7